NAME-
DROPPING

....

FROM F. D. R. ON

BOOKS BY

JOHN KENNETH GALBRAITH

[A PARTIAL LISTING]

American Capitalism:
The Concept of Countervailing Power

A Theory of Price Control

The Great Crash, 1929

The Affluent Society

The Scotch

The New Industrial State

The Triumph

Indian Painting
(with Mohinder Singh Randhawa)

Ambassador's Journal

Economics and the Public Purpose

Money: Whence It Came, Where It Went

Annals of an Abiding Liberal

A Life in Our Times

The Anatomy of Power

A View from the Stands

Economics in Perspective:
A Critical History

A Tenured Professor

The Culture of Contentment

A Journey Through Economic Time:
A Firsthand View

The Good Society:
The Humane Agenda

Name-Dropping: From F.D.R. On

John Kenneth Galbraith

NAME-DROPPING

FROM F. D. R. ON

HOUGHTON MIFFLIN COMPANY
Boston • New York
1999

For information about permission to reproduce selections from this book, write to Permissions, Houghton Mifflin Company, 215 Park Avenue South, New York, New York 10003.

Library of Congress Cataloging-in-Publication Data

Galbraith, John Kenneth, date.
Name-dropping : from F.D.R. on / by John Kenneth Galbraith.
p. cm.
ISBN 0-395-82288-2
1. Politicians — United States — Biography Anecdotes.
2. Galbraith, John Kenneth, 1908– — Friends and associates — Biography Anecdotes. 3. United States — Politics and government — 1945–1989 Anecdotes. 4. United States — Politics and government — 1989– Anecdotes. 5. Politicians — Biography Anecdotes. I. Title.
E747.G28 1999
973.9'092'2 — dc21 99-20070 CIP

Printed in the United States of America

Book design by Robert Overholtzer

QUM 10 9 8 7 6 5 4 3 2 1

The photographs in this book are from the author's personal collection, gathered over many decades. In most cases, the copyright holders could not be identified. If necessary, the publisher will add the proper credit in future printings.

For Catherine Atwater Galbraith
who shared in all this

CONTENTS

NAME-DROPPING

....

FROM F. D. R. ON

1

On Name-Dropping

Books, like those who write them, have an unplanned life of their own. The very act of writing has a controlling role. When I started this book, I intended to describe the political personality—the personal and public traits that, as I saw them, allowed the great leaders of our century to influence or dominate the political scene. There are still elements of this intention in the pages that follow. But it faded as a central purpose.

Instead, as the work proceeded, there was more interest for the author, as I trust there will be for the reader, in how the great political figures appeared to their contemporaries, of whom I was one. What did I recall of personal encounters or public association with Franklin Roosevelt, Eleanor, the Kennedys, Nehru and others? Such recollections took over, but with them came a certain risk.

Reminiscence and anecdote, as they tell of one's meetings with the great or the prominent, are an established form of self-enhancement. They make known that one was there. This is not my purpose; my aim is to inform and perhaps, on occasion, to entertain. The risk, nonetheless, exists that critics who are less than tolerant may suggest that I am indulging in name-dropping. Hence the title of the book and that of this chapter; nothing so disarms a prosecutor as a prior confession of guilt.

Not all that follows concerns the political figures of my time. I frequently digress to write of my own experience and of responsibilities accorded me. This tells something of those of whom I speak. Not exceptionally in writing of this kind, it may well tell more of the author.

Here also is an occasional event or personal encounter of which I have told before. For this I do not apologize. All education and all worthwhile writing is, in some measure, a recapture of the already known.

Much of this book—most, in fact—is centered on now-distant times; an important part dates to the first half of the century that is now drawing to a close. It was with the events of this period and the people that I was involved. I now read of, and from time to time encounter, the influential men and women of the present day. It is for others to tell of them; this I do not, in all cases, regret.

There will be question less as to those who have been selected for recollection and celebration here than as to those omitted or discussed only briefly. The reason is not far to seek; it is whether or not I was there and have something to add. On one or two occasions I met Dwight D. Eisenhower; he was and remains one of the underestimated Presidents of our time. A Republican, he accepted the great social legacy of Franklin D. Roosevelt and the twenty-year New Deal era and made it an integral part of American life. F.D.R. initiated, Truman continued, Eisenhower confirmed. He also left the deathless and death-defying warning as to the military-industrial complex. But when I have said this of Ike, I have very little else to say.

As to another major figure, one exactly of my generation, there is a similar problem. Ronald Reagan and I were fellow founders of Americans for Democratic Action, once and still a dominant liberal voice in the land. Ronnie, as he was known, left us when his screen career diminished and he began giving well-paid lectures on, as it was then denoted, the free enterprise system. His regression, we always said, was not from any commitment to newly acquired belief; it was only for the money. On his later career there was nothing of which I had firsthand knowledge. This, I do slightly regret, for Ronald Reagan was the first wholly uninhibited Keynesian President—eager public spending to provide economic stimulation and employment, all financed by large public borrowing, with the resulting budget deficit. However, there was a dark side, to which Keynes would have reacted adversely: the spending was for extensively unneeded armaments.

With Jimmy Carter, whom I first met in Georgia and saw on later occasions, I had only a distant association. It was his special tragedy that, while Ronald Reagan succeeded with the economic policies his party had so long opposed, Jimmy Carter was taken to defeat by those conservatives had long urged. His highly reputable economists, in pursuit of economic virtue, accepted that a President seeking re-election could survive inflation attacked only by its traditional and painful remedies: high interest rates, economic stagnation, unemployment. It was a triumph of rigorous economic orthodoxy; ignored only was Jimmy Carter's all-but-certain fate.

One of my closest and certainly one of my most admired friends in politics over many years has been George McGovern, presidential candidate in 1972 against Richard Nixon. I had a small role in his selection as a candidate and a not insignificant one in his defeat. At the Democratic Convention that year, as a leader of the Massachusetts delegation, I vetoed his first choice for Vice President, Kevin White, the Mayor of Boston. I did not think I could win state support for his nomination because, among other things, White had endorsed McGovern's opponent in the primary. There would be an unseemly row on the floor. McGovern went on to Tom Eagleton, who, it soon became known, had once had some modest, wholly curable psychiatric problems. Unwisely, George dropped him from the ticket and then was involved in an embarrassing search for a substitute. In consequence, his campaign had a very bad start. He should have ignored my advice. I haven't told here of George McGovern perhaps because, again, I have little to add, perhaps more because I prefer to write about those with whom my association was less disastrous.

Also passed over with McGovern, but for a very different reason, is Richard Nixon. In 1942, in the tense months after Pearl Harbor, he served in the Office of Price Administration as an attorney on rubber-tire rationing, of which I was then in charge. He drafted my letters, but I did not, as I recall, ever meet him. I became fully aware of his existence and character only with his crusade against Communism and Alger Hiss. Later when his enemies' list became known, my name was present, adorned, according to my recollection, with two checkmarks. In one of his taped and reluctantly released conversations in the White House, he dignified me as the leading enemy of good public process in our time. But, to repeat, I never met him, so Richard Nixon is not here.

I once contemplated a chapter in this book on Winston Churchill and Charles de Gaulle. Not in recent times, not perhaps ever, have two politicians accepted greater risks with greater ultimate success. How grim and dim the prospect in 1940; how enormous our debt to their intransigent stand. During my wartime years in Washington, Churchill was especially a presence; one thought of him, more even perhaps than of F.D.R., as the guiding military force of the war. I did meet both Churchill and de Gaulle but only after the war was over and for no deeply operative purpose. To have made anything of these encounters would, indeed, have been name-dropping.

A more serious matter is the very few women—only Eleanor Roosevelt and Jacqueline Kennedy—present in these accounts. That, very simply, is because, for most of the period here covered, women were not visible in the political world. The concern here is with high office; this was the virtually exclusive domain, the preserve, of men. Among presidential wives some did step forward. In her husband's presidency Nancy Reagan was an evident force; with her, not surprisingly, I had no personal acquaintance.

John F. Kennedy, in a conversation of which I have told on other occasions, once raised with me the question of women in politics. He advanced what I thought the deeply retrograde thesis that women were naturally lacking in political talent. He asked me to name some outstandingly successful women politicians. I responded with Eleanor Roosevelt. He agreed and asked for another. I was troubled for the moment and, in some desperation, proposed Elizabeth I. Kennedy laughed scornfully and said, "Now you have only one left, Maggie Smith." Margaret Chase Smith, pioneer woman senator from Maine, was not—here we differed—a favorite of his.

Were Kennedy now alive, he would not be making the point; women are still underrepresented in politics, but the change in the last thirty-five years strongly affirms their political aptitude. Alas, it came too late for this volume. And there is yet to be a woman President.

I turn now to Franklin Roosevelt, the first and in many ways the greatest of those I encountered over a lifetime. And the one, more than incidentally, who accorded me the most responsibility. It was no slight matter to have control over all the prices of all things sold in the United States. And briefly over consumer rationing as well. My role in the Office of Price Administration was my principal association with F.D.R., but I also observed his leadership in the New Deal and, more generally, in the war, and of this I will tell as well.

2

Franklin D. Roosevelt, I

THE NEW DEAL

PRESIDENTS of the United States were not in my early youth in Canada a large factor in my life. As an elementary school student during World War I, I did hear of Woodrow Wilson, of his non-role and then his role in supporting the Allies and therewith the Canadian troops who were so deeply involved in that ghastly encounter. This, rather often, was a subject of adult conversation. In subsequent years, however, I was only dimly aware of the occupants of the White House. I do recall—it was obviously in November of 1920—my father picking up the *Toronto Globe,* the bible of the committed Liberals of Ontario, and observing that a man named Harding had been elected President. It seemed not very important.

By 1932, I was a graduate student at the University of California at Berkeley, and my mood was still one of detachment; in any case, as a Canadian, I could not vote in that year's elections. As I've elsewhere told, my first, if somewhat distant, encounter with a President was with Herbert Hoover and a speech he delivered that autumn from the back of a railway car at an obscure station in Oakland. Unfortunately, it was adjacent to what was then called a Hooverville, this one a dump of large sewer pipe boarded up at the ends to provide exceptionally low-cost housing. The residents, having no other occupation, lined up early for the oration; they cheered loudly when he told them that the Depression was over and that, in principle, they were prosperous again.

One reason for my general lack of interest, as was true for others of my generation, was that, in those days, both of the established parties and their candidates seemed deeply irrelevant. In the dominant belief of my fellow graduate students and some of the California faculty, the problem was not political choice; it was the system. F.D.R., like Hoover, was a hopelessly outdated prospect. Those who understood and accepted the full reality of the time were socialists or, quite frequently, Communists. Caution and a slight uncertainty as to whether, as a recent refugee from farm life in Canada, I would be accepted kept me away from the latter. In a famous expression, Marx had referred to the idiocy of rural life, and that was my history. (I might note that I told of all this years later in a magazine article, and still later I learned from my FBI file that two particularly ardent economics professors, worried that I was about to become the president of the American Economic Association, had brought my questionable past to the urgent attention of

J. Edgar Hoover. Hoover was grateful in response but took no action.)

Franklin Roosevelt's victory in the autumn of 1932, like the campaign before it, did not seem politically or socially decisive. It was good to see Herbert Hoover go; it was of no great significance that Roosevelt would take his place. There had been little that foretold the wide-ranging action that was soon to come. Conservatives, then as they would be now, were reassured by F.D.R.'s promise, made in a major and later much-regretted speech in Pittsburgh, to balance the budget. Then, as now, there seemed nothing that would so adequately and conveniently exclude costly and otherwise unwelcome social action.

So when the 1932 election results were known, I had no feeling of excitement and not very much interest. My greatest sense of change came from a young colleague who verged on ecstasy at the thought that alcoholic beverages would soon be legal. No more need to resort to the chemistry lab.

Nor, with more others than the history tells, was I aroused by F.D.R.'s inaugural address. That we had "nothing to fear but fear itself," his best-remembered words, was palpably untrue; there was much, much else to fear. The speech, we thought, only echoed the previous rhetoric from other voices in Washington who had urged that nothing was needed but a restoration of confidence in the economic system.

My attitude, like that of others, changed radically in

the ensuing weeks as the Roosevelt program, the New Deal, took shape. Something, indeed, *was* happening: there was action on a large scale. Over the Christmas holidays of 1933, I made an expensive trip to Washington for a firsthand view. I saw much of the city and only a little of the effort, but I returned to California as an accepted expert on F.D.R. and what he planned for the country. Recently hired as a teacher at the University of California at Berkeley, and dispatched to teach undergraduates at Davis (then an extension of the Berkeley campus), I became the local authority on the now-diverse and impressive Roosevelt initiatives. I lectured not only to the deeply indifferent undergraduates but to interested faculty members and the larger community.

My commitment to Roosevelt was further deepened the following summer, when, on my way from Berkeley to Harvard, with the intervening summer months free, I was recruited into the Department of Agriculture.

The USDA was, at the time, a focal point of New Deal thought and action. Assembled there, under Henry Agard Wallace, the Secretary, were all the significant radicals of the age, from Rexford G. Tugwell to Adlai Stevenson and on to Alger Hiss and some avowed Communists. I was still a Canadian, but, as I've also often told, I was not asked as to my citizenship; I had only to affirm that I was a Democrat. My assignment, which came from Tugwell with the sanction of F.D.R., was to consider whether the millions of acres of farm and forest land made tax delinquent by the Depression should be taken into the public domain. Much of northern Michigan, Wisconsin, Minnesota and extensive areas in the Old South would thus become federal property—enough for several wonderful national parks. The modest recompense involved would be a windfall for the local governments that would thus be relieved of this redundant real estate. Alas, my recommendation that we should so proceed received no serious attention.

I did not meet F.D.R. that summer. He was, nonetheless, the major, indeed the dominant, figure in my life. So he was for all the others I knew. We talked of him all day and every night; there was no other topic of conversation. We were responding to a central part of the Roosevelt political personality, one that still deserves emphasis. He was a man of intelligence and a deep sense of social responsibility, but he was also without a controlling personal ideology, social belief, of his own. That meant that he was available to be persuaded; he was open to any well-stated solution to the great and painful problems of the time. In further consequence, scores, even hundreds, could feel that he was theirs to convince and to provide the needed action. No one ever said, "You can't sell the President on that"; it was possible that you could. And this was not the feeling in Washington and the government alone. It was generally believed that the administration of Franklin D. Roosevelt gave or appeared to give the possibility of hearing and response. There had been nothing quite like it before; there has been nothing like it since. In Washington and out in the country determined citizens felt that they could be a part of history.

For most Americans the Depression was a far more painful experience than the war that followed. World War II brought fear, death and sorrow to a relatively small minority; the Depression, in contrast, brought deprivation and deep insecurity to the many; in fact, to most. Save for those who marched, sailed or flew against the Germans and the Japanese (and by no means all of them), the war was a good time for Americans. Jobs were available, pay was good, the general standard of living was raised. People were dedicated to a high purpose. This was far from true of the grim existence during the accurately designated Great Depression. It focused attention on the dismal economic problem, and here there was no fixed, no decisive, opinion. Roosevelt was, to repeat, free of ideological commitment. This meant that all who were socially motivated could, in imagination, if not in reality, influence the future course of economic policy. This seeming opportunity increased as the months passed, for the President was also known to be deeply bored with economics.

Lauchlin Currie, my fellow Canadian, a brilliant economist and an early disciple of John Maynard Keynes, whose work he had somewhat anticipated, was the first economic adviser, so named, assigned to the White House. On one occasion of which he told me later, he was called to Warm Springs to brief the President on the economic outlook and needed action, but he was able to do so only on the train back to Washington. He had brought papers and charts, which he marshaled in series before the President. After a glance at each, Roosevelt turned back in silence to the window. Had he had a confident and thus a dominant view as to economic policy, he would have made Currie a mere instrument of the presidential will. Instead, Currie could feel, as did many, that it was for him to express that will and F.D.R. would agree. That was very much my feeling when I was later given major wartime responsibility. All of those around F.D.R. admired and loved him because his actions were partly ours or could seem to be partly ours. Thus, admiring ourselves, we admired him.

The matter must not be carried too far. Roosevelt was impatient with detail, certainly with technicality; these especially he delegated. But larger goals were on his mind. The United States he saw as a vast estate extending out from his family home at Hyde Park, New York. For this he had responsibility, and particularly for the citizens and workers thereon. Sometimes this sense of responsibility manifested itself in a direct and visible way. Early on in his first administration he made a trip to the Plains states—Kansas, Nebraska, the Dakotas. There, his thoughts turned to the bleakness of the land; how much more appealing it would be were there trees. He returned to Washington to propose an extensive tree-planting program—a "shelterbelt"—and it went forward.

This was the reaction of a great landlord, an obvious step to improve appearance and property values, a benign action for the tenantry.

Not that all was peaceful and harmonious on the estate. There was an especially bitter conflict that even to this day calls for explanation. The programs of the New Deal had, it was clear, two separate effects. FERA, the Federal Emergency Relief Administration; WPA, the Works Progress Administration; PWA, the Public Works Administration; the Triple A, the Agricultural Adjustment Administration, all provided jobs or income to the needful. So, as to income, did Social Security—old age pensions and unemployment compensation—when it appeared midway in the 1930s. The humane effect of these programs was not questioned, and certainly not by those they helped.

There was also their macroeconomic effect, which extended to the entire economy. They provided an addition to purchasing power—in economic terminology, an increased flow of aggregate demand—and this, in turn, was of economic benefit to the economy as a whole. It was an obvious answer to the diminishing or stagnant demand that, above all, characterized the Depression. Directly benefited were business firms and business leaders.

Of yet more immediate assistance to business was the National Recovery Administration, the NRA, which brought companies together to write the famous codes that allowed the fixing of prices and an arrest of the deadly price- and wage-cutting that was one of the most clearly adverse effects of the Depression. A warm response from the companies and corporations affected was naturally expected. A few executives did, in fact, approve, but most were in dedicated and increasingly articulate opposition. Roosevelt was considered not a friend of business but a relentless enemy.

The reason for this, which has never been fully understood, lies in the two motivating forces that exist in any economic system: one is the desire for money; the other is the need for prestige. The pursuit of money—income—is widely accepted. But for the business community, prestige is also deeply important and something not to be shared. The only acceptable economic policy is one that accords front rank to the corporate executive or the financier. An active government, like Roosevelt's, all too obviously challenges the basis for business esteem and self-esteem. Better to suffer some loss of income, even accept recession or depression, than to see this prestige—this right of leadership—impaired or invaded.

In the Roosevelt years there was general puzzlement as to why businessmen so strongly resisted public action that had the effect of stabilizing and enhancing their own financial return. They did so because their eminence, their sense of who was in charge, was challenged. This is still the case; business attitudes today are deeply influenced by the need to be thought the decisive force in economic life.

Once the New Deal measures were in place, the business community rallied strongly in opposition to Roosevelt. The American Liberty League, made up of especially articulate entrepreneurs, came into being as early as 1934. Roosevelt was the object of their attack; he had appropriated the seeming economic dominance that properly belonged to them. They were being helped, but that was not the important thing.

Ideology, needless to say, had a role. The free enterprise system needed to be defended, and the business community saw itself as its defender. There were memories of distant college instruction and the message of read and unread books by journalists, free-lance philosophers and rigorously orthodox academicians. This was the approved motivation. But deeper and more powerful was the sense of lost position, of prestige passing to Washington from New York, Pittsburgh and Detroit. It was that, as much as ideology, which mobilized the businessmen against Franklin D. Roosevelt. Paradoxically, however, their antagonism became a considerable source of strength for him as well.

In all politics one needs an enemy and preferably an incompetent, misguided or socially adverse one. The business community served Roosevelt wonderfully in this regard. No one could doubt the prominence that its leaders had enjoyed under Herbert Hoover; now they expected the same position, even though the dismal aftermath of his administration was known by all. Roosevelt could not have had a more desirable opposition. This he recognized; increasingly he accepted and publicly identified big business and its spokesmen as his enemy. His subordinates joined the battle with special enthusiasm.

The business leaders who were aligned against Roosevelt were also, needless to say, in alliance with the Republican Party. The political result was spectacular; Alfred M. Landon, Governor of Kansas, a man of distinctly moderate view and later my friend and host in Kansas, was saddled in the election of 1936 with the burden of business support, and he went down to a resounding defeat. Only Maine and Vermont, the latter eventually to become the most liberal state in the Union, gave him their votes.

In the years following Roosevelt's 1936 triumph there was another factor enhancing his popularity: not only was there the much-enjoyed combat he led, there was the even deeper joy one felt in being with the winner. This, too, is a powerful force in political life. I digress to recall a personal example.

By the 1936 election I was fully involved in presidential politics—the journey of enlightenment from that word of the Harding victory was complete. I spoke in Roosevelt's support, argued vigorously the Roosevelt cause. The latter was not easy in my academic generation at Harvard; one had some difficulty finding anyone on the other side. I was untroubled by the fact that I was still not a citizen, could not myself vote, but after some urging from Massachusetts politicians whom I came to know, I took steps to overcome this last barrier to being a fully committed Roosevelt man. In the company of a large group of fellow immigrants, almost all of them responding to a regulation that limited relief payments to citizens, I was called one morning before a federal judge. He told us we would be sworn in in alphabetical order. Referring then to a paper that had been given him by my political co-workers, he called for "Galbraith." The next to be summoned was, as I remember, Aaron or Abraham or possibly Anderson.

From the autumn of 1937 until the autumn of 1938, I took time away from the United States to be part of the highly committed community assembled around John Maynard Keynes at the University of Cambridge. And to see the evolving welfare state in Sweden and have a look at Hitler's Germany and Mussolini's Italy. At the end of the year I returned home to resume teaching at Harvard and to accept a major assignment in the Roosevelt White House.

It was to direct a study of the principal New Deal programs, specifically the Works Progress Administration and the Public Works Administration, and assess what they had done for employment and for the economy as a whole. My employer was a presidential agency, the National Planning Board, later, in an exercise of semantic caution, renamed the National Resources Planning Board. The honorary chair of its largest committee was Frederic A. Delano, the uncle of the President; one of its members was Henry S. Dennison, head of the paper products company of that name in Framingham, Massachusetts, the most liberal of American businessmen. Dennison offered his employees unemployment compensation and retirement pensions long before the government decree requiring them. Immediately in charge was Charles W. Eliot II, a Roosevelt favorite and member of a family with which I have had a long association.[1]

While directing this study, I had my first and one of my relatively rare personal encounters with the President. He was in the Oval Office, seated behind the great desk. The light from the office windows flowed over his shoulders; he reached out to bid us welcome and tell of his active support for what we were doing. He had an aspect of health and energy, and it was an impression that remained with me—indeed, became more vivid—in the years to come. The case was the same for the public at large. Unable to walk, in the common reference a hopeless cripple, he was nonetheless seen and heard as a man of normal vitality and vigor. Perhaps some of this was owing to what was, for F.D.R., the fortunate absence of television—his voice on the radio was untroubled and strong. Perhaps also there was a certain reticence on the part of the press to disclose his disability. But mostly it was personality, as it is called. He presented himself as a man without weakness, and this was how the country

1. His younger brother, Thomas H. Eliot, author of the Social Security Act of 1935, later a congressman from Massachusetts and later still the greatly effective Chancellor of Washington University in St. Louis, was my lifetime friend, as has been his wife, Lois.

and the world perceived him. So certainly we all regarded him until that dismal April day when news came of his sudden death. No one had any knowledge of, or had given any thought to, his possible ill health. Our belief in his invulnerability was complete, and that belief was shared by the whole country.

I return to the study I was directing. It involved a contentious matter with two exceptionally combative advocates. The WPA, headed by Harry Hopkins, was created for the good of the workers; the test of success was not the task performed or the thing built but the employment provided. This, the WPA saw as the greatest need in those years.

The PWA, supervised by Harold Ickes, the Secretary of the Interior, was geared to the end product—the thing built. The test was the resulting post office or bridge. To my pleasure, and adding appreciably to my sense of importance, I was lobbied ardently for my support by both WPA and PWA officials. So was F.D.R., who, characteristically, left the matter to the combatants. From the effort to persuade him came one of the most memorable of the Roosevelt stories of the time.

While F.D.R. was at Warm Springs, Harry Hopkins came in one morning to make the case for the WPA. What could be so important as the employment offered? Jobs were the great need; to provide them was what the New Deal was all about. It was the WPA that must be accorded priority in public emphasis—and whatever money was available.

Listening, F.D.R. nodded agreement. "Harry," he said, "you are perfectly right."

That afternoon Ickes came to the cottage at Warm Springs to make the case for public works. The valid test should be the utility and general excellence of the thing built. It was by this that the whole effort would be judged; it was on this that the reputation of the administration would depend. To public works, attention and money should go.

The President considered carefully the Ickes case. He said, "Harold, you are perfectly right."

Eleanor had been in the room during both of the visits. When Ickes left, she took over. "I don't understand you, Franklin. Harry comes in and makes his case, and you say he is perfectly right. Then Harold comes in and makes the opposite case, and you tell him *he* is perfectly right."

To which the reply was "Eleanor, *you* are perfectly right."

When eventually completed, my study,[2] not surprisingly, gave strong approval to the employment effort and the borrowing this required. The latter—a firm endorsement of Keynesian doctrine—was still a delicate matter, for it was at the margin of respectability or slightly beyond. A couple of high officials in the old-line government departments who read the manuscript were seriously disturbed. They conceded that the contents were part of the Roosevelt agenda; the latter, however, was to be tolerated, not avowed. I held my ground.

2. *The Economic Effects of the Federal Public Works Expenditure, 1933–1938* (National Resources Planning Board, Public Works Committee, 1940).

The report was published in 1940, just as all interest, my own included, had turned to the defense program and the threat of war. It attracted almost no attention. I doubt that F.D.R. ever saw it.

With the war in Europe came a new challenge to Roosevelt and the Roosevelt polity, one not fully recorded in the histories of the time. In 1940, it became clear that public support must be rallied to provide aid to Britain and to counter the less visible threat from Japan. Then in 1941 came Pearl Harbor and Hitler's incredible declaration of war on the United States. The government would now be the guiding force in the vast production of war matériel. This brought back into focus, in a different and in some respects an intensified form, the problem of Roosevelt's relations with the business community and its leadership. There were highly evident enemies abroad, but there were for F.D.R. less evident enemies here at home. He was engaged on two fronts. The myth, the nearly universal myth, is of a country united by war. The reality is of a continuing and sometimes damaging political conflict centered on Washington. Of this I can speak as a young but deeply engaged combatant.

3

Franklin D. Roosevelt, II

THE WAR IN WASHINGTON

THE WILLINGNESS, almost pleasure, with which Roosevelt took on the leaders of the great business community in the New Deal years does not mean that he was in any way indifferent to political opposition. His inaction on civil rights, which did not receive any legislative attention during his three and more terms, was a clear example. The Northern black, now Afro-American, community responded warmly to his concern for employment and to the work and relief programs; the Democratic vote in the great urban centers still reflects that commitment. But Southern blacks, who were equally poor, were largely neglected. White Southerners in the House and Senate would be automatically averse to civil rights legislation, and their support was not only needed but essential. This, Roosevelt accepted. Accordingly, no real change was even contemplated, some mostly symbolic efforts for sharecroppers by early Roosevelt radicals, as they were regarded, in the Department of Agriculture alone excepted.

In the autumn of 1940, I had my first significant association with F.D.R. as a politician: I was recruited for his speech-writing team in the election of that year. It was my first professional experience in this art; one should always start at the top. We operated from a small suite of offices in the Commerce Department. This was a public building, and all of us there involved were on the public payroll. Nothing was paid for by campaign funds. Political standards were relaxed in those days. Our speech drafts went over to higher-level talent at the White House. I did not see Roosevelt personally during the campaign, a large and agreeable White House reception apart. We found out how much of what we had written had survived when we heard the President speak on the radio. We listened with some attention.

The most memorable contribution came from a highly competent economist on our staff, G. Griffith Johnson. Roosevelt was by now encountering Republican criticism for the poor condition of our defense forces. Noting who had fought to curtail defense appropriations in the Congress, Johnson came up with three wonderfully euphonious Republican names: the House Minority Leader from Massachusetts, Joseph W. Martin, Jr.; the once-noted advertising man, now a member of Congress, Bruce Barton; and F.D.R.'s own congressman, the articulate reactionary, Hamilton Fish. These names went into a speech Roosevelt was to deliver at Madison Square Garden. As he stood at the podium, the President chuckled for a moment, asked who had led the opposition to defense appropriations, and then rolled out, "Martin—Barton—and Fish." The response was ecstatic. A night or two later in Boston he asked who had voted against adequate appropriations for agriculture. It was not exactly a rural assembly; nonetheless, there was a galvanic reply: "Martin, Barton and Fish." I have been involved over the years with a certain amount of speech writing, including some for presidential candidates. This speech for F.D.R. remains first in my memory.

I return to the larger scene.

In 1940 and 1941, Roosevelt moved with caution on military intervention in Europe. Isolationist sentiment, the America First syndrome, was strong in the land, and it had F.D.R.'s deep respect. However, beginning in 1940, with the fall of France, he took action to strengthen the American armed forces, which were then roughly on a par with those of Portugal. The industrial base for the production of arms and munitions was also broadened, and support was accorded the British through the seemingly benign destroyers-for-bases deal and then Lend-Lease.

On serious military involvement, however, Roosevelt was cautious; he led only as he could be assured of followers. What eventually committed America to the war was not his leadership but the unbelievably insane actions of the enemy. The Japanese attack on Pearl Harbor in December 1941 was superbly designed to arouse the most reluctant America Firster. Then, within hours—hours spent in Washington in deep concern that all attention and military resources would now be diverted to the Pacific and even that the attack had been a tactic designed to achieve that result—came Hitler's declaration of war on the United States. I have difficulty describing the relief that was felt upon hearing of this monumental insanity.

Years later, at an interrogation of Joachim von Ribbentrop, the Nazi Foreign Minister, in a high-level prison in Luxembourg, he was asked the reason for this departure from elementary intelligence. He replied that Germany had been bound by the terms of its treaty with Japan and Italy. A young bilingual aide who was handling the translation asked on his own, "Why was that particular treaty the first one you decided to keep?" No American ever solved so grave a problem for F.D.R. as Hitler did in December 1941.

Remaining, however, was Roosevelt's most pressing wartime dilemma: his already-mentioned difficult relations with the business community. The conflict of the thirties now continued in a new and even more aggravated form. The circumstances were dismayingly simple: business firms run by businessmen would have the responsibility for war production, but to bring businessmen to Washington, especially those of adequate public reputation, and place them in positions of real authority was to surrender control of the economy to a deeply committed opposition. Successful corporate leaders could not be expected to disembark in the nation's capital and become Roosevelt acolytes. Or even be minimally neutral. The Germans and the Japanese were the enemies; on that, all agreed. But Roosevelt remained a clear and present threat. The country had to be defended against attacks from abroad; in the view of F.D.R.'s numerous opponents in the business world, it had also to be saved from him and the New Deal.

To all generalizations there are exceptions. Some business leaders did give their qualified and sometimes their unqualified support to Roosevelt. These included corporate lawyers, investment bankers and company executives: Henry Stimson, lawyer; Averell Harriman, investment banker and onetime railroad magnate; Will Clayton, the great Texas cotton merchant and reformed member of the American Liberty League. (He left it when his wife started matching all his contributions to Republicans with equal amounts to F.D.R.) Also the young Nelson Rockefeller and Jesse Jones, Texas financier and longtime head of the Reconstruction Finance Corporation, the RFC. Lurking always in the neighborhood of the White House was the highly available Bernard Baruch of World War I industrial mobilization fame, who was in constant search of a second coming. All accepted Roosevelt. All, in one way or another, gave their best to the war effort. They were, however, in sharp contrast with the business community as a whole. It is not without interest that the business leaders who aligned themselves with Roosevelt are today remembered, while the names of those who resisted are lost not only to fame but to memory. In fact, many did resist; war did not justify surrender to the government and to Roosevelt. Belief and principle had their claim.

Roosevelt's strategy was to give seeming authority to the business community but the decisive power to the military, to civil servants and especially to Roosevelt men—the New Dealers. The impression of business participation and support would come from business notables of good reputation and little practical effect and from business-led bureaucracies of great size and visibility and no really decisive role, the War Production Board being the principal example.

There were undoubted problems with this design. To accord seeming power without delegating real power involved, at first, the creation of a series of war mobilization agencies that offered a vision of action. Each was intended to appear more effective than the last, and each soon revealed its too obvious ineffectiveness. There were the National Defense Advisory Commission, the NDAC; the Office of Production Management, the OPM; the Supply Priorities and Allocation Board, the SPAB. All emerged within the space of a few months; all were visibly incompetent; all were soon replaced. Each had been created to give the business community the appearance without the reality of authority.

The War Production Board was more impressive than its predecessors; it attracted a huge influx of business talent, some with previous Washington experience in trade associations or as lobbyists. On its leadership and role a later word.

Of central importance was the enlistment of business leaders of high visibility and, in practice, low competence. Faced with the inordinately complex task at hand, they were passive and often helpless. Thus, for example, William Knudsen, former head of General Motors, was dominant in the National Defense Advisory Commission and, with labor leader Sidney Hillman, one of the two heads of the Office of Production Management. He brought an undoubted air of authority; here, one could be sure, was a corporate tycoon. Unfortunately, he had no useful impression of what industrial mobilization and procurement required. At a tense assembly of officials to which I was summoned on the night of Pearl Harbor, he was asked, in sonorous tones, "Bill, what are the marching orders for tomorrow?" After some thought, he replied that there would still be a serious shortage of copper. He had no idea as to what was to be done about it. Since he could not be sacked—that would have been a visible blow at business—he was eventually made a general. To the all but certain military advantage of the Republic, he commanded no troops.

Edward R. Stettinius, Jr., equally visible in the defense effort, also had admirable business connections—U.S. Steel and General Motors. With prematurely white hair, singular beauty and nearly total incompetence, he was as impressive as Knudsen, as inadequate but, in his progress through the government hierarchy, more fortunate. After his non-service on the National Defense Advisory Commission, he was made head of the Lend-Lease program. This was an extraordinarily undemanding post, since all its significant duties were handled by the American, British and Russian agencies immediately responsible for war procurement. (I was briefly assigned to Lend-Lease when my tenure in charge of price control came to an end, and I remember it as the most depressing exercise in unrelieved idleness I have ever known.)

Later, when Roosevelt, tiring of the inconvenient obduracy of Secretary of State Cordell Hull, was determined to have a committed cipher in his place, he chose Stettinius for the post. My friend George W. Ball once told with pleasure of the way Ed showed him how he had modernized the ancient and then somewhat senile State Department: he had had his office painted. In the closing days of the war, when, as previously mentioned, our forces took Joachim von Ribbentrop into custody, he had on him two letters urging the British to join with the Germans against the advancing Russians. They were addressed to *Vincent* Churchill. Ball, who was there, thought that had we been in a similar situation, Stettinius would have written to *Albert* Hitler. Stettinius's later years were unforgiving and sad; he went on from the government to a disastrous business venture in Africa and an early death.

To be head of the War Production Board, ostensibly the most significant wartime mobilization agency, Roosevelt appointed Donald M. Nelson. He was an amiable, undoctrinaire figure, who, as a former retailer, was more in touch with public and political attitudes than were the true Roosevelt adversaries. Under him the Board's enormous staff, recruited from industry and, as noted, its trade associations, continued to see its responsibility as containing Roosevelt and the New Deal. Of this intention there was constant, if low-key, expression. Otherwise, there was little effective action. The miracle of wartime arms production must be credited not to the WPB but to the direct procurement operations of the Army and Navy and such devoted figures as General Brehon B. Somervell (who, in a civilian interval in his earlier career, had managed relief operations for the New Deal in New York); General Lucius Clay, thereafter brilliantly in charge of Occupation affairs in Germany; and a group of younger men, including Robert McNamara, of later distinction and acute public controversy. But more of the American production miracle was simply the response of an underutilized and underemployed economy to the wonders of an unlimited demand for its products.

The Roosevelt strategy also called for reliable supporters —Roosevelt men—being present in the wartime agencies and dominant in those of key public significance. On the War Production Board under Nelson, Simon Kuznets, the father of the Gross National Product and, after Keynes, the greatest economic figure of his time, and his disciple Robert Nathan, one of the country's truly effective liberal economists and statisticians, dominated production planning: the setting of production goals and the monitoring of associated performance. Nathan, compulsively articulate, was intensely unpopular with his business contemporaries, and when he was eventually drafted into the Army, it was to their great pleasure. Leon Henderson, the most thoroughly committed Roosevelt man of all, was given a seat on the War Production Board. He was also given responsibility for price control, rent control and consumer rationing. The resulting agency, the Office of Price Administration, was a solid Roosevelt instrument, but, more important, it was one of the two wartime organizations with the greatest public reach; its actions would affect every citizen. The other was the Treasury, which was in charge of taxation and had considerable influence on public expenditure. This too was in solidly New Deal hands under Henry Morgenthau, Jr., the Secretary of the Treasury, and Harry Dexter White, his Director of Monetary Research and later a major target of the Red hunt. Although it was little mentioned at the time, the OPA and the Treasury were, from a political point of view, the centrally important agencies during the war.

In the spring of 1941, I was given charge of price control and later, for a time, rationing under Henderson. As rationing became a separate entity, I remained the price czar. I too was, unqualifiedly, a Roosevelt man. On this history a personal word.

*

In the summer of 1940, a few days after the fall of France, I had been summoned to Washington to work with Leon Henderson on monitoring price movements and to formulate price policy as defense mobilization got under way. It was then a non-job; prices were still stable, even depressed. I was moved on to the more substantive task of overseeing the location of the new defense plants to ensure, among other things, that they were not all heavily concentrated (as in World War I) in the industrial Northeast. In that role I became involved that autumn in the sharpest controversy of the time as between the Roosevelt and the anti-Roosevelt forces in the defense effort. It was a textbook case on the tensions of the pre-war era and, indeed, was later so described and published.[1]

The issue was the location of an ammonium nitrate plant, producing, as the recent tragedy in Oklahoma City made widely known, a substance that could be both a farm fertilizer and a formidable explosive. In World War I such a plant, and a dam to provide the power for making the constituent ammonia, had been located on the Tennessee River, at Muscle Shoals, Alabama. It never became fully operational, but its possible use as a source of cheap fertilizer attracted strong rural support and was for many years bitterly controversial. To its opponents in the chemical and fertilizer industries, it was a clear case of socialism. On coming to office, Roosevelt resolved the

1. Harold Stein, *Public Administration and Policy Development: A Case Book* (New York: Harcourt, Brace, 1952).

conflict over Muscle Shoals in a manner more unacceptable to its enemies than could have ever been imagined. Muscle Shoals became the launching pad for the whole Tennessee Valley Authority. Never was a political battle so lost and so won.

The TVA had been created primarily to build dams and produce electricity; it had never produced fertilizer, even though farmers had long wanted a public fertilizer plant. Now, with the war and the need for the explosive, it was felt that this final step should be taken. The opposition from the chemical and fertilizer industry and from business in general was venomous: here was Franklin D. Roosevelt at his worst.

Using the authority of the early National Defense Advisory Commission, which was responsible for plant location, Stettinius and Knudsen, who represented the business interests on the Commission, tentatively approved two private plants, one for Du Pont, one for Allied Chemical, both safely away from the Tennessee Valley. Muscle Shoals and the TVA were left in abeyance. Those of us who had supported their development took the issue to F.D.R. He replied that he would like to see all three plants under the management of the TVA, a wonderfully typical example of his manipulative skill. Almost immediately, by a unanimous vote of the NDAC, Muscle Shoals was approved; better that the Authority have charge of one plant rather than three. So great was the ensuing industry backlash that Knudsen and Stettinius then asked that the record be changed to show that they had voted in opposition. Nonetheless, we had won. I went for a short vacation to reflect on how much better it was to be on the Roosevelt side. Since I had had the primary responsibility in this matter, my role and my success therein led to my being given my next assignment, by most calculations one of the major wartime posts.

Some months before Pearl Harbor, in the spring of 1941, I was given charge of price-control operations, including the shaping of the requisite organization to carry them out. The next year, in the winter after Pearl Harbor, my task was broadened, as I've said, to include the first steps in the establishment of a rationing system. I have never been sure as to how much Roosevelt knew of my role; I did receive approving word from him when, in 1942, the General Maximum Price Regulation (General Max), which was considerably my initiative, achieved something very close to effective price stability. Before then, however, I had had an experience that revealed clearly the Roosevelt sensitivity to political considerations.

Not too long after Pearl Harbor it became evident that the Japanese, relatively unimpeded, were going south through Southeast Asia, where they would overrun the rubber plantations in Malaya and Indonesia. There would be no more natural rubber, and our manufacture of synthetic rubber was still some difficult months distant.

On short notice, and with the support of equally concerned colleagues, I froze all retail and wholesale stocks of rubber tires and put into effect a stringent rationing system. If a tire was to be sold, it had to be essential for the national defense, for medical personnel, for public transportation or for other similarly compelling need. The test of such need was severe. On the implementation of the edict, a message came to me personally from Roosevelt asking what congenital idiot had supposed that ministers of the Gospel were not essential. Particularly, he asked, had I never heard of Southern Baptists and their political impact? In a day or two, ministers became essential.

I mention two other incidents that tell something of the mood of the time. In addition to the rationing, and following the advice of a committee led by Bernard Baruch and Harvard President James B. Conant, we set a speed limit of thirty-five miles an hour on the nation's highways. This, it had been determined, would mean a substantial reduction in the wear on tires. We asked the governors of the states to enforce the measure; all agreed except Coke R. Stevenson, the exceptionally regressive Governor of Texas. It fell to me to persuade him. I called, and I still remember his response: "Doctor, here in the state of Texas when you drive *thirty*-five miles an hour, you don' get there!"

In this same period a fire in Fall River, Massachusetts, destroyed between 15,850 and 20,000 tons of natural rubber. This precious commodity had been taken over by the Rubber Reserve Corporation, an offspring of the Reconstruction Finance Corporation that, as earlier mentioned, was headed by Jesse Jones. Another Texan, Uncle Jesse, as he was called, was both the most prominent and the most durable of the emissaries to Roosevelt from the business community. His basic faith, however, like that of others of his generation, had not been altered. Reporters hearing of the fire rushed to get his comment. He replied that they needn't worry; the rubber was fully insured.

There were limits beyond which even the most devout Roosevelt man could not go. In the autumn of 1942, Leon Henderson resigned under heavy assault from industry. We in the OPA were well loved for our efforts to protect the civilian living standard against inflation but not by those whose prices were kept level in the process. There was an especially adverse reaction when, rather unwisely, we went beyond price control to try to regulate the quality of the products we were pricing. It had been feared that pea soup might be reduced to one pea per can; rayon hose, now replacing silk, might be good for only one outing. With an undoubted excess of caution, we set standards for these products. The response from industry was especially severe. To this day I think of canned goods and rayon with unease.

Suffering the full force of the attack, I was called almost daily before one congressional committee or another to answer for our actions, especially the orders regulating quality but, more generally, for those keeping prices below what sellers knew they could receive. A leading food industry magazine put permanently on its editorial masthead the words GALBRAITH MUST GO.

There was also dissension within the agency. Prentiss Brown, a pleasantly ineffective former senator from Michigan and a Roosevelt man, had taken Henderson's place, and he brought in an obscure and moderately obtuse public relations expert from his home state to improve our image. The latter saw me as the principal reason it was so bad. I responded by telling him, in one of the more restrained expressions at my command, that he could mind his own damn business. It was too much; there were limits to what could be suffered even from a Roosevelt supporter. I was moved to the already-mentioned non-employment at the Lend-Lease Administration. My critic went back to Michigan. It was still better to be a Roosevelt man.

The hours and days of idleness bore heavily upon me —as noted, the worst days I have ever spent. I went to my draft board to volunteer for the Army; after what must have been a slightly informal inquiry, I was told that my height—six feet eight-and-a-half inches—disqualified me; in World War I the trenches had not been that deep. I returned briefly to civilian life as an editor of *Fortune* magazine, then a superior center of literary excellence. Archibald MacLeish, Eric Hodgins, Dwight Macdonald, James Agee, all reflected the reluctant decision of Henry Robinson Luce, the founding father, that it was better to employ liberals who could write—even devout Roosevelt supporters—than conservatives who could not. Then, in the spring of 1945, as the war was winding to a close, I was called again to a Roosevelt enterprise; the presidential need was always a command. He had asked Secretary of War Henry Stimson to make a study as to what our bombing of Germany had accomplished. All previous reports had come from the Air Force generals; they had learned what they knew from aerial photography and were not thought to have minimized their own achievements. Thus was established the United States Strategic Bombing Survey, USSBS—popularly known as Usbus. I was put in charge of determining the overall economic effects of the air attacks. My duties included the interrogation of the high Nazis as they came into custody and, on one major occasion, as I will later tell, even before that.

The day before I was to leave for Washington and on to London and Germany, we were to have a small farewell dinner in New York. I had taken my children to school that day wearing the uniform of a civilian colonel, and I met some Afro-American GI's, who saluted me impressively. I discovered that I did not quite know how to respond. Later that afternoon a friend called and told us to listen to the radio; it was being reported that Roosevelt was dead.

So he was. Our dinner with fellow Roosevelt hands and supporters was an occasion for both shock and grief—shock because we had supposed Roosevelt was forever, grief because he had always truly had our love. Worst of all was the feeling that we suddenly had been severed from the future. What for the country and for ourselves would now come? And how could Roosevelt's sudden passage from life be explained?

Roosevelt had, in fact, been dangerously ill for months, but, as I've earlier told, the impression he conveyed was always one of health, strength and command. Perhaps even had I and those similarly involved heard of his illness, we would have passed it over; Roosevelt was a permanent fixture in our lives.

The day after the news, on the train to Washington, I shared the trip with Nelson Rockefeller. He too, as already noted, was a Roosevelt man, and he was charged at the time with our relations with Latin America. We discussed what the loss of the President meant for the country and, more specifically, what it meant for us. Even for a Rockefeller, it was now the great unknown.

In April 1995, Catherine Galbraith and I made our way to Warm Springs, Georgia; it was fifty years after the President's death, an anniversary remembrance. The landscape and the Roosevelt cottage came rather as a surprise. What once was farming country is now heavily forested. The room where the President suffered the stroke is small, rather dimly lit. I had always imagined something light and spacious; I had to adjust years of thought. There were only three or four of us from the Roosevelt years surviving and in attendance. In his speech President Clinton took notice on the platform of Arthur Schlesinger, Jr., who had been a young participant in the Roosevelt days and who became his pre-eminent historian, and of me in the front row of the audience. He urged that the country respond to the principles still being enunciated by these Roosevelt hands. That was personally rewarding. We applauded warmly. So even more was the memory of the greatest political personality of the century—the leader against the Great Depression and the most necessary and unforgiving of wars.

4

Eleanor Roosevelt

ONE DAY in the early 1940s, when I was engaged in price-control operations in Washington, I had suddenly to go to my dentist. A broken tooth. I waited outside his office, and presently a pleasant-looking, well-proportioned woman came out. She said, "The doctor told me you were here and that you were at Princeton." This I acknowledged; I had been teaching there.

"My husband used to be there too," she said, adding an agreeable word about that distinguished academy. When I got to the dentist's chair, he asked if I had spoken with her, and when I inquired who she was, he told me, "Mrs. Woodrow Wilson."

It is widely held that, in the last months of the Wilson Administration, after the President had suffered his stroke, Mrs. Wilson was, effectively, the Chief Executive. Yet she was little known, then or later. Had it been Eleanor Roosevelt in the dentist's office, failure to recognize her would not have been possible for even the most primitive observer. Not before Mrs. Roosevelt, and quite possibly not since, has a President's wife enjoyed such recognition, been so clearly seen to have her own role in life and her own public agenda. One went to her for help with the President; one went far more for her personal guidance and assistance. The President functioned from the White House and Hyde Park, as did Eleanor, but she operated also from an apartment in mid-Manhattan. As F.D.R. had the radio and his Fireside Chats, she had her newspaper column, "My Day." She traveled extensively and comprehensively; a famous *New Yorker* cartoon of the time showed two coal miners deep below ground amidst dirt and gloom. A figure is approaching through the dark, and one of them is saying, "For gosh sakes, here comes Mrs. Roosevelt."

Doubtless there were such references, but I do not remember Eleanor Roosevelt ever being called the First Lady. Certainly I never called her that. The title is not wholly decent. It does not recognize independent intelligence and aptitude but is simply a consequence of marriage. A succession of recent presidential wives—Jacqueline Kennedy, Lady Bird Johnson, Nancy Reagan and, one cannot doubt, Hillary Rodham Clinton—have lived their own lives and had their own agendas. So, but in more marked degree than any of them, did Eleanor Roosevelt. We did not think of her as a presidential artifact, nor did she so think of herself. She existed independent of the President; it was even a noticed and accepted fact that they did not travel much together. She was present on social occasions at the White House, but no one imagined that these were a large part of her life. We thought of Eleanor Roosevelt as someone who, but for the accident of history and the prevailing constraints of gender, could have been President in her own right. This last she herself did not assert, but neither, one judged, did she doubt it.

Eleanor Roosevelt was a trifle above medium height, and she gave an impression of strength. Her face was firm, with the beauty of evident intelligence. Her dress was agreeably utilitarian; the effect was always of tweed. It would have been hard to think of her in a truly fashionable evening gown; she was not a precursor of Jacqueline Kennedy. One saw her as a dominant guide, colleague and co-worker of the President, not just as his wife.

In 1947, at the founding meeting of Americans for Democratic Action, Eleanor, with Franklin, Jr., sat toward the front of the assembly. She wore a dark suit and a small dark hat. She was surrounded by the liberal legatees of the New Deal, but there was no question as to who quietly, effectively, was the most important person there. So on all such occasions. It was Eleanor Roosevelt who was in the room—and others.

I first met Mrs. Roosevelt in the autumn of 1940 at a large, impersonal White House reception for recently recruited defense personnel. Many of us were young; it was a major event. An even younger friend of ours and his wife owned only an ancient Ford, so they drove to a parking space a few blocks from the White House, where they hailed a taxi to take them to the official entrance. They had to arrive in suitable style; the old car would not do. On the slowly moving reception line, I came to Eleanor Roosevelt. She grasped my hand, welcomed me to my work in Washington.

A meeting of more consequence was in her New York apartment that same autumn. Donald Comer of Birmingham, Alabama, a major Southern textile manufacturer—the Comer-Avondale Mills—was a man of liberal disposition and action, race relations possibly apart. Not surprisingly, he shared the already-mentioned view that a substantial proportion of the expanding defense industry should not be located in the Northeast. Were it placed instead in the South, it would help to modernize the still primitive rural plantation economy and its narrow parochial politics. Pursuing this purpose, he proposed that we seek the support of Mrs. Roosevelt. She had a natural interest in such matters.

An appointment was made; we took the train to New York, the subway from the station, and late in the afternoon came into her apartment. Her quarters had a well-used appearance; the upholstery was somber and more than slightly worn. Through the fading light we told her of why we came, what wartime industry would do for the South. Anticipating our meeting, she had already given the matter thought; this she made clear, telling us as much as we told her. She would see that the President was properly persuaded. All was brief and to the point.

She did as promised. Word of the President's agreement came from the White House. Important defense plants went to the Southern states. I allow myself a digression to tell of one of the consequences.

As a major part of this effort a large shell-machinery factory was proposed for Gadsden, Alabama. (It was to Gadsden that the rubber-tire companies had earlier gone to escape the unions, enjoying the full cooperation of the local leaders in their flight.) Sidney Hillman, the labor representative on the National Defense Advisory Commission, which briefly had a controlling authority in such matters, vetoed the proposal. We could not have a plant in a town he could not visit himself.

Joseph—always called Joe—Starnes, congressman from the district and an articulate reactionary of the time, came to the support of our plan. With so many jobs and so much money at stake, Joe said that he could not stand on principle. He persuaded the town fathers to say that union organizers would be accepted; they would be treated with true Southern hospitality. This reference to Southern hospitality Hillman at first rejected as an escape clause, but with the help of others his acceptance was obtained.

So the war came to Gadsden and so did factory workers and so did unions. So did political action, and Joe Starnes was the natural target. At the next election, he was thrown out. Mrs. Roosevelt, if she knew, would have liked the result of her effective support.

Later, as the war was ending, I encountered Joe Starnes. I refrained from asking him how he felt now about standing on principle.

*

Eleanor Roosevelt's life was not traumatically changed by her husband's death nor, I think, by the fact that Lucy Mercer Rutherfurd, his longtime friend and mistress, was with him when he died. She accepted life—and death—as they came. She had always been independent, and so she remained. Lucy Mercer was a small detail. On her a word.

The fact that the President had a mistress was first revealed to me, as to many others, in the book by Jonathan Daniels.[1] This I read only a few days before Truman Capote's famous Black and White Ball, a huge assembly of the mostly liberal intelligentsia in 1966. My dinner partner that evening at the home of Jean Stein, the well-loved New York literary hostess, was Alice Longworth, daughter of Theodore Roosevelt and the senior member of the Roosevelt clan. Inevitably, I asked her about the book and about Lucy Mercer. She snapped back, "It means nothing. Everyone knows that Franklin was paralyzed from the waist down."

In the years following the President's death, Eleanor became even more intensely active. A delegate to the United Nations, she was also chair of the UN Commission on Human Rights, took the lead in obtaining the Universal Declaration of Human Rights and gave firm attention and support to the civil rights movement in the United States. While F.D.R. still lived, she had

1. *The Time Between the Wars: Armistice to Pearl Harbor* (Garden City, N.Y.: Doubleday, 1966).

once been sharply criticized for consorting with Afro-American leaders, having her picture taken with them. Now she was free from political constraints.

As Adlai Stevenson emerged on the political scene, Eleanor was strongly attracted to him, as he was to her. She responded to his political style and especially to his interest in, and commitment to, foreign policy and the United Nations. In his 1952 campaign, along with other of his speech writers, I was in communication with her. Early on we sent a delegation to brief her, telling of our political strategy, so far as one existed, and of the issues we intended to emphasize in the candidate's speeches. Our emissaries came back in a mild state of shock. Her instructions as to what Stevenson should say and do were far more specific, far more acute, than anything they had to suggest.

Through the 1950s Eleanor Roosevelt was one of the less visible but most important people in the Democratic Party. She did not actively campaign; she did speak and write. She was a force behind the scenes. It would have been impossible for any Democrat to run for high office without her support, and certainly not if he had had her opposition. This was something on which we did not reflect; it was assumed. On foreign policy she had a strongly negative reaction to Secretary of State John Foster Dulles and his Cold War militants and militarists. On domestic matters she was the custodian of the whole New Deal legacy, with a special concern for the poor. This being so, those she did not support so far as the

Democratic Party was concerned were consigned to the political wilderness. This latter fact played a part in the closest, most significant encounter I had with her.

Late in the decade, in addition to her writing and other activities, she came abreast of the times with a television program emanating from Brandeis University in Waltham, Massachusetts. In consequence, she was often in Cambridge; we met and lunched or dined together, usually in the company of equally committed liberals. I was now supporting John F. Kennedy for the Democratic nomination and for the presidency. It was important that he have her endorsement.

Nothing could have run more radically against her political instinct. The Kennedys were known to be a close-knit family; J.F.K.'s father, Joseph P. Kennedy, Sr., had broken sharply with Roosevelt on the decisive issue of wartime support to Britain when he was Ambassador to the Court of St. James's. He believed we were backing a loser. Eleanor's adverse feeling was very deep; it extended automatically to Joe's son.

I urged that the sins of the fathers should not be visited on their offspring. It was to no effect. Then, in the last days of 1959, while appearing on her television program, I proposed that she relent at least to the extent of having J.F.K. as her guest on a future broadcast. Rather to my surprise, she agreed. As so often, commitment to an audience overcame political differences.

The program a short time later went very well; both

participants were interesting, even mildly eloquent. I was in a state of warm self-congratulation. The press had rallied around, and after the show Mrs. Roosevelt was asked the most pressing question of the time. Kennedy had, that very day, stated his intention of seeking the Democratic nomination. Would she support him? Leaving no one in doubt, she said she *certainly* would not. That, needless to say, was the principal news that came out of the meeting. Kennedy was not impressed by my role as a political strategist. At dinner that night he reflected on the deeply ingrained political incompetence of professors.

By the summer of 1960, however, when J.F.K. had become better known, had shown he was clearly his own person, and the Kennedy charm had become evident, she relented. She came out in full support. The election being very close, it is quite possible that she made the difference.

During the campaign, along with other guidance, she said that someone, meaning me, should see Bernard Baruch and ask for a contribution. His was still an influential voice; by his own assertion, he was an important Roosevelt man. But unless he had invested money in a candidate, his allegiance was never certain. He needed to be asked for financial support. Accordingly, the next time I was in New York, I made a date with him, went up Fifth Avenue, listened to a couple of hours of self-serving reminiscences, got a contribution and was assured that he was with us. Nixon had already ruined his own case by asking Baruch, then ninety, to head a "Senior Citizens' Committee" during the campaign.

"Me, a senior citizen? What does he take me for?"

When Eleanor and I next met, I told her with some self-approval of my errand and my success. She responded, as I recall, "Oh, Bernie. What a man! By 1945, he and my husband had broken off their relationship. He was, as ever, using the White House for personal advertisement, and that had finally created a rift. But it didn't matter to Bernie. When he heard of the President's death, he came right over to Warm Springs. He was with us there. He came with us on the train to Washington. He was with us for the state funeral. He came on the train with us to Hyde Park. He was there at the family funeral. And there were several times, Ken, when I thought he was going to get into the coffin with Franklin." This was Eleanor Roosevelt: the whiplash tongue.

Early in the morning of November 8, 1962, while I was Ambassador to India, word came to New Delhi that Eleanor Roosevelt had died. She was seventy-eight. My thoughts ranged back over the years; no one in public life had I so respected, in fact, if often at a distance, so loved. With both Roosevelts gone, it was, indeed, the end of time.

I immediately had the flag at the embassy placed at half-mast. Later, my deputy, a distinguished Foreign Service officer, came in to tell me that this could be done only on orders from Washington; the flag should go up to full-staff. I lost my temper and told him in an obscene way what he could do to himself. Then immediately I regretted my explosion and my language and apologized. But I also had the flag stay where it was. At noon a telegram came from the White House, ordering that the flags on all embassies around the world be flown at half-mast.

A question remains. How important was Eleanor Roosevelt in directing the course of the socially and politically most decisive administration of the century? One should not exaggerate her contribution. The decisive role was that of F.D.R.; and it, in turn, was shaped by the two major events of the time, the Great Depression and World War II. Eleanor Roosevelt was a subordinate player. She provided access to the President, as here told; she gave reassurance to all those who sought or needed humane action. No one was a greater champion of the poor. She was active both at home and abroad in human rights. Far more than her husband, she was a creature of conscience and was so recognized. Still, the response to the primal force of history came from F.D.R.

5

Albert Speer

THE ESSENTIAL ENEMY

DURING WORLD WAR II and in the years following, one of the most elaborately contrived public figures in modern history was Albert Speer, Hitler's armaments chief and, effectively, the director of Germany's war economy. His eminence was, to no small extent, his own achievement, but, in substantial measure, it was also the creation of the opposing Allies. Then, and especially in the history of the era, there was need for a seemingly worthy enemy, and Speer, more than anyone else, has so served. To this I will return.

Albert Speer began as Hitler's architect. He charmed the Fuehrer with architectural and landscape proposals of the kind by which tyrants, over the centuries, have sought to be remembered—great scale and great cost often as substitutes for grace or beauty. Then, in the war years, as already noted, Speer was put in charge of German arms production and, more broadly, the German war economy. Most of the high Nazis—Paul Joseph Goebbels, Heinrich Himmler, even Hermann Goering—have now faded from memory and to some extent from the historical record. Martin Bormann, Hitler's closest intimate, whose possible escape from Germany in the last days of Berlin was a matter of great speculation, is forgotten. Over Hitler's admirals and generals, not excluding Admiral Karl Doenitz, his unfortunate successor in office, the curtain has also come down. The name of Albert Speer almost alone remains. Television documentaries have told of his life, personality and accomplishments; so have books, including in particular his own. Not only is he the best-remembered figure of the Third Reich, Hitler apart, but, more remarkably, he is the one least burdened by its crimes. This is all the more extraordinary because he was far from competent in the position in which he served. And he was also deeply involved in some of the more ghastly enterprises of the Nazi regime.

Albert Speer, without experience in such matters, was put in charge of arms production in early 1942, when Fritz Todt, his predecessor in office, was killed in an airplane crash. After more than two years of war the German manufacture of arms was still at modest levels; in the major categories it was below that of Britain. Under Speer in the following years it increased but at a slow and uncertain pace. Conversion from consumers' goods to armaments was cautious; women were not called to work in the factories. Domestic servants remained available, and more were brought in from the Ukraine. Night shifts were exceptional. Only the repair and rehabilitation of bombed plant and transport were carried out with dispatch, as the Allied air forces, to their distress, were later to learn. Arms production generally did not reach its peak until late 1944, as the war was coming to an end.

None of this was denied by Speer; in interrogation after the German surrender he admitted to the flaws in his management. He told of his grave discontent at seeing in *Life* magazine the myriad of American women at work in the shipyards and aircraft plants. Germany had no Rosie the Riveter. Still, he took no action to enlist women, initiate night shifts or otherwise enhance military production. Instead, slave labor was recruited and put to work. Those so compelled toiled, not surprisingly, without enthusiasm. The cruelty and suffering to which they were subject Speer did not deny. He did attribute much to Fritz Sauckel, the Nazi labor plenipotentiary and war criminal. Under interrogation Sauckel was more forthright. Of Speer he said, "There is a man you should hang." Sauckel was hanged.[1]

During the war there was little public mention of Albert Speer in the United States and Britain. Goering, Goebbels and Himmler were much more celebrated. Speer's name did appear prominently in the files of the

1. Cf. *A Life in Our Times* (Boston: Houghton Mifflin, 1981), p. 212.

OSS (the Office of Strategic Services, predecessor of the CIA), and by the end of the war he was well known by those professionally associated with wartime intelligence. But not by others. I did not, that I recall, hear of him until I arrived in Germany in the spring of 1945.

My association with Speer began in Flensburg on the Danish border a few days after the war ended. The Doenitz government in succession to Hitler was functioning there while the Allies contemplated the surrender of a government that had already twice—once in Rheims to the Western Allies and once in Berlin to the Soviets—surrendered. As earlier told, I was a director of the United States Strategic Bombing Survey, assigned to see what the bombing raids on the Reich had actually, in fact as opposed to hope, accomplished.

The interrogation of Speer took place over several days in the magnificent and undamaged Schloss Glucksburg, not far from Flensburg, where he was quartered. Sessions began each day at noon, for Speer was serving as Minister of Economics and Production in the Doenitz Cabinet, and the latter assembled for largely ceremonial purposes each morning at eleven. Speer, proudly showing his knowledge of American idiom, said, "It is what you call Grade B Warner Brothers."

I participated with George Ball in a later interrogation of Speer at Kransberg Castle in central Germany, the prison under British auspices that carried the code-name Dustbin. The American camp for high Nazis at the spa town of Mondorf-les-Bains in Luxembourg, where we also conducted interrogations, had the code-name Ash-can. There was effort at humor even in those grim days.

Albert Speer was tall, slender, attractive in appearance and easy and responsive in speech. Under interrogation, he gave the impression of having somehow anticipated every question he was asked, as quite possibly he had. On our formal responsibility, that of assessing the effect of the air attacks, he provided competent answers re-inforced by a large store of statistical material, which, thoughtfully, he had brought with him out of Berlin and stored temporarily in the vault of a bank in Hamburg. This, of course, we obtained.

It was on his relations with Hitler and the Nazi hier-archy, and the implied responsibility of any high Nazi for the concentration camps, the death camps and, as mentioned above, for slave labor, that he showed his careful prior thought. This would also be evident in later months leading on to, and at, the Nuremberg trial. Ours was a splendid preview of his design for survival and eventual rehabilitation.

He did not deny responsibility or guilt. That, he knew, would be the effort of some of his erstwhile colleagues; it would identify him with the wrong people and, in any case, would be wholly implausible. Nor did he detach himself from the death camps: "I was told of things hap-pening there which I should never see." On Nazi war op-erations in general he presented himself as a participant but as one who was unique—he was there, but he was apart. He had not come from the Nazi Party apparatus or the active role of a confirmed Nazi. It was Hitler who had brought him in; he was an acolyte. He was devoted to the Fuehrer but not to the Nazi hierarchy. It was his strong sense of a personal association that caused him to fly in to Berlin to say farewell to Hitler on almost his last night, when the Soviet forces were only a few kilometers from the bunker and Hitler only hours from suicide. Perhaps, a less persuasive matter, it also prevented him from dropping poison gas into the ventilating system of Hitler's bunker in the last days of the war. This he said he contemplated but in the end refrained from doing.

Not being part of the Nazi machine, he could look with detachment and even some contempt on those who were. In later weeks I took part in interrogations of Hermann Goering, Joachim von Ribbentrop and Field Marshal Wilhelm Keitel and Colonel General Alfred Jodl, the last two Hitler's immediate military staff and described by Speer as Hitler's nodding donkeys. I also saw something of Julius Streicher and Dr. Robert Ley, both militant Nazi voices. It was not difficult to see how different they were from Speer. They were indeed—I do not exaggerate—an incredible collection of often deranged incompetents. Speer knew that this would be the judgment of history and moved to make clear his separate standing. This led him to point out the extraordinary role of alcohol and drugs as the Nazi regime was coming to an end. Goering, it was known, was a drug addict; in prison when we questioned him that summer, he was still in bed suffering the pangs of withdrawal enforced by his American captors. Others were experiencing the similar, if less dramatic, effects of sudden sobriety. This prospect Speer had emphasized at Flensburg: "When the history of the Third Reich is written, it will be said that it drowned in a sea of alcohol...In those last months, I was always dealing with drunk men."

This, of course, is not wholly remarkable. Over time, not a few decisions of marked eccentricity have been taken by politicians and soldiers who were less than sober. Historians rarely mention that fact; on alcohol and the decisions it affects, there is an accepted convention of silence. The clear case of the high Nazis in Germany was seized upon and used by Speer in differentiating himself.

Albert Speer's life was spared; he was sentenced to twenty years in Spandau prison. There, his attention turned, successfully, to saving and enhancing his reputation. He arranged with an old associate to have some part of his record suppressed, notably that concerning his expulsion of Jews from Berlin. He let his self-discipline be known. Measuring the distance each day on a map in his cell, he set about walking around the world. Upon his eventual release, he was still a man of evident vigor. He wrote favorably and, as circumstance required, apologetically of his career. He traveled and gave interviews, including one to George Ball, my fellow interrogator in

Flensburg. When he died, he was, on balance, a well-regarded figure.[2]

The ability of Speer to save and enhance his reputation (and indeed his life) came from his undoubted success in separating himself from the other Nazis, his fellow criminals. But that was not all; there was a further circumstance in his favor, one that is little recognized and calls for emphasis. It is the need in war and in later history to give stature to the enemy. During wartime this is necessary to inspire the best efforts of the troops. "We face a formidable foe; prepare for the worst." No military axiom is more frequently, more sententiously, repeated than "Do not underestimate the enemy."

Then, when the war is over, those who won do not wish the public perception of the strength and intelligence of the now-defeated foe to be diminished. This would lessen their own wartime achievement and detract from the victory. Only the defeat of a worthy opponent secures the history of the conflict for the victors, and they, indeed, write the history.

This was especially important in World War II. In that war, following the long military tradition of their country, German soldiers, sailors and airmen were courageous and effective. The German political and military

2. As already suggested, he had developed his plan for survival, perhaps even recovery of public esteem, by the time the war ended. George Ball and I told of this in an article in *Life* magazine in 1945, and to my pleasure, our view has been cited and affirmed in the recently published and admirably researched *The Good Nazi: The Life and Lies of Albert Speer,* by the Dutch author Dan van der Vat (Boston: Houghton Mifflin, 1997).

leadership was, however, of manifest incompetence. How otherwise could one explain the already-mentioned declaration of war on the United States, this as the German armies faced the terrible test of the Russian winter? Or the devastating order to hold fast at Stalingrad? Or the belief that something could be accomplished by the Battle of the Bulge? The Nazi war record was replete with stupidity. In the long history of military ineptitude few can rival Hitler for strategic error. Or those who advised him.

Given this incompetence and its recognition as such, it was necessary for the Allies, especially when peace came, to look for any evidence that the enemy had been worth the wartime effort and sacrifice. The top Nazis and their supporting generals simply would not so serve; their inadequacy, their mistakes, were too evident. Hitler was a special disaster. There remained Speer; he alone was capable of giving the Nazi leadership an aspect of intelligence. Accordingly, he was seized upon as a formidable figure, far better than the drug addict Hermann Goering; the relentlessly loose-lipped Paul Joseph Goebbels; the visibly dim-witted Joachim von Ribbentrop or the reckless maniac, Hitler himself. Or their military advisers, again the nodding donkeys. Speer achieved much of the distinction he came to enjoy because he, virtually alone among those at the top, seemed a worthy foe.

I close with a word on how this appeared to a soldier, an early airman, of the time.

My civilian rank and duty in Germany in 1945 allowed

me an airplane but not necessarily a safe one. The Canadian C-64 I was assigned was underpowered for the load it carried. It was flown by an alert former fighter pilot, a Belgian, who was even more disturbed than I by the inadequacy of our transport. He eventually escaped a crash landing when, to my pleasure, I was not aboard.

On the day of which I tell, we were flying down the Moselle after an interrogation session with the high Nazis at Ashcan in Luxembourg. Knowing that he was deeply and intelligently curious about these men, I had arranged with the head warden to have him taken on the morning rounds of the prisoners—Goering, deep in drug withdrawal, and the other decayed, physically dilapidated and largely depressing congregation. He returned sad and close to tears. Finally he told me the reason for his grief: "Who would have thought we were fighting the greatest war in history against that bunch of jerks?" I too reflected on this as the vineyards passed serenely below us. Albert Speer, whom the pilot had not seen, would have slightly redeemed the situation, as he later did for so many others.

6

Harry Truman—and After

In the winter of 1946, I was called back to Washington from my writing and editorial duties at *Fortune* to have charge of German, Austrian, Japanese and South Korean economic affairs so far as these reached to, or emanated from, the State Department. The responsibility was not overwhelming. My path of authority ran through the State hierarchy to the Joint Chiefs of Staff at the Pentagon and then by uncertain passage to Berlin and Tokyo. I did develop a close and wholly admiring relationship with General Lucius Clay, who, as earlier noted, directed American Occupation affairs in Germany; I drafted the later, somewhat famous Stuttgart speech given by Secretary of State James F. Byrnes, which, among other things, returned control over the German economy to the Germans. I do not believe that General Douglas MacArthur in Japan was aware of my existence. President Harry Truman was only modestly so. He asked me, along with General John H. Hilldring, then retired from the Pentagon to the State Department, to add to our other duties keeping a close eye on the policy for Israel, or what was so to be called. He did not trust the old and socially eminent State Department hands. It was not only that they were pro-Arab; they were also—the fact must be faced—anti-Semitic. I supported the Jewish state—then the reference—at the meetings of the CORC, the Coordinating Committee, which consisted of the heads of the several departmental offices. One of my fellow members was Alger Hiss, who, on all issues, went comfortably along with the Establishment.

Harry Truman was of unassuming presence, deeply conscious, perhaps overly conscious, of his meager formal preparation for the post he had so suddenly assumed. Roosevelt always felt that he was meant to be President; Truman was astonished that the office had descended on him, of all people. His qualifications, as others came to recognize, were not slight. He had an active mind and superb common sense. He was willing to assume responsibility—"The buck stops here"—and he had a good sense of humor. He read prodigiously and informed himself well on policy issues. He had political experience and keen political instincts. The latter were soon evident in the way, after taking over from F.D.R., he went on to carry his party to the unexpected and unexampled triumph in the election of 1948.

In the accepted view of the American polity, all parts of the country, all social classes, have an equal voice. So the textbooks teach; so every good citizen believes. Before Harry Truman the reality in this century was different: the United States government was, to a singular degree, the preserve of an Eastern elite. There were exceptions, to be sure, but both Roosevelts, Woodrow Wilson, Calvin Coolidge, were all from the Eastern seaboard. Herbert Hoover thought of himself as belonging to New York, and there he retired. The most durable spokesman of the Republican Party, Robert Taft, was from Ohio, but he came from a family once based firmly in southern Vermont. There was little mention of this predominant Eastern orientation, and emphasis on it should not be carried too far. There were effective figures from the Plains and the West: the La Follettes from Wisconsin, the Wallaces from Iowa, Hiram Johnson from California and the diminishing Midwestern voice of William Jennings Bryan. The point, nonetheless, remains: the federal government was largely controlled by the Eastern states and by the graduates of their educational institutions.

With this regional dominance, Harry Truman was the decisive break. He had served and been advanced by the Pendergast machine in Kansas City; there had been no one else with quite such a background. Before he became Vice President, Truman had been a moderately effective senator. The wartime committee he headed—the Senate Special Committee to Investigate the National Defense Program, known universally as the Truman Committee—had done a very useful job in tracking down excessive and mendacious payments and other fraud in the war contracts. It had created a limited but valuable atmosphere of fear. Nonetheless, Harry Truman, the product of a small Midwestern town and a politically dubious urban machine, was something entirely new in the Oval Office. No one felt this more acutely than he did himself. It was generally assumed that he was there only to finish out the Roosevelt term.

That thought was widely shared in 1948. I saw it in an engaging way on election night. I was by then back in Cambridge reinventing my academic career, and I did not, beyond voting, take part in the campaign that autumn. Nor did most others of the Cambridge and Boston liberal community; Truman was not their kind of candidate.

After the voting booths had closed, members of our convocation gathered in considerable numbers at the ample and welcoming residence of Arthur Schlesinger, Jr., there to share news of the inevitable defeat. All assumed that the Democrats and their accidental leader would go down before Thomas E. Dewey. (One of the few bright moments of the immediately preceding weeks had come from the campaigning of Harold Ickes, Roosevelt's Secretary of the Interior and the most articulate liberal of the day. Taking note of the Republican candidate's reticence on pressing any politically troublesome issues, Ickes had vigorously attacked "Thomas *Elusive* Dewey, the Candidate in Sneakers.")

Before going to the gathering, I listened to the radio, and it seemed that things were not in line with expectations. I commented in a casual way to my wife that the dapper Republican standard bearer might well be suffering some doubts: "I think Thomas E. Dewey may well be shitting in his blue serge pants."

At the Schlesinger house, we found our political coreligionists gathered for drinks and the usual intense conversation. Reacting to my earlier misgivings, I went right to the radio and was joined there by Barbara Kerr, a friend, distinguished journalist and undoubted liberal, who was, however, paid to write editorials for the *Boston Traveler,* a very conservative newspaper, now long and deservedly extinct. Presently she went upstairs, came down with her coat and fur neckpiece and joined me again as I listened to the news. Then she got up to go. I asked her why she was leaving. She replied, "That editorial on 'Return to Sanity' isn't going to run. I've got to do a 'No Mandate for Socialism.'"

The next day the most famous of all political photographs became available. It was of Harry Truman joyfully holding up an early edition of the *Chicago Daily Tribune* with the blackest of all headlines: DEWEY DEFEATS TRUMAN.

Although it was not recognized at the time, the Truman political personality had its own very special appeal. Roosevelt, as I've earlier observed, saw himself as the overseer of the great national estate, a landlord who assumed responsibility for his tenants. Truman, in contrast, saw himself as of the tenantry; he was the kind of person the people knew. The popular reference was to

Harry Truman, not to President Truman or the President. The advantage this conferred was something those of us attending that election night soirée in Cambridge did not understand; and, with deeply adverse effect, would not understand any better when Adlai Stevenson became the candidate. Strongly, even affectionately, favored by the scholarly elite across the land, Stevenson did not identify with the population and the voters as a whole. He was weak where Truman was strong. To this I will return.

Truman was in one other way in contrast with Roosevelt. In politics and public statement, F.D.R. always left himself room for later decision; it was often far from clear what he intended to do. Truman invited no such question, left no such doubt. He was routinely described as "plainspoken"; this was a widely welcomed change.

My strongest personal memory of Harry Truman comes from the late 1950s, when, the Democrats being out of office, a special advisory organization of scholars and still-eager job aspirants was created in connection with the Democratic National Committee—the Democratic Advisory Council. Dean Acheson and I were co-chairmen, he for foreign policy and, as he saw it, the need for an aggressive stand in the Cold War; I for domestic policy and the steps on from the New Deal. We met at regular intervals; after prolonged and often repetitive discussion we issued statements on needed public action. When Acheson had finished his formal presentations, Averell Harriman would always comment, "Well, Dean, I see you've declared war on the Russians again." I would then agree. Acheson was as much at odds with my liberal views on domestic policy as I with his aggressive ones on the Cold War.

One day Truman came to a meeting. For the occasion the press and radio were allowed in; they arrived with enthusiasm and much equipment. It was Harry Truman, after all. I was presiding that day, and Truman joined me at the head of the conference table. He had just come down from New Haven, where, as a Chubb Fellow at Yale, he had been a visiting scholar. I have told before of the ensuing exchange:[1]

"How, Mr. President, were things at Yale?"

"You know, Professor, every university is like every other. Students all Republicans, faculty all Democrats." Then he continued: "People who talk about the differences remind me of once out in Independence when I went down to the courtroom to fill in the time.

"We had a prosecuting attorney in those days who was really pretty good. One day he had a fellow up for rape, and he opened up by saying, 'Your Honor, I intend to show that this man had sexual intercourse with this poor unwilling woman. And, Your Honor, I will show that he was also guilty of fornication with this reluctant young woman.

"That kinda surprised the judge. Since we were pretty

1. Cf. *A Life in Our Times* (Boston: Houghton Mifflin, 1981), pp. 360–361.

informal out there, he said, 'Bill, I don't understand. What's the difference between the two?' And Bill said, 'Judge, I have to admit I've tried them both, and they're pretty much the same!' That's how universities are in my experience. Different names, otherwise the same."

At this moment Charles S. Murphy, once counsel to Truman in the White House, later an Under Secretary of Agriculture and Chairman of the Civil Aeronautics Board, hurried to the front of the table behind which we were sitting. He held up a piece of paper that said, THE MIKE IS LIVE.

Beads of perspiration came to my forehead. Not to Truman's. He looked at the message and dismissed it. "They can't use that. It's too dirty."

Truman was a professional politician; that is how, rightly, he is remembered. The true professional assesses the future with competence and deals confidently with the prospect. This, Truman did. He correctly sensed the need, as peace came, for bringing major economic resources to the support of the Western European states. This alone would help ensure an orderly and democratic recovery from the wartime devastation and postwar chaos. In consequence, a total in the range of a hundred billion present-day dollars passed through the Economic Cooperation Administration to be used in overseeing the Marshall Plan, an enormous sum in a time when populations and economies were smaller and budget appropriations were much more modest. (It is in wonderful and also dismaying contrast with what was later made available by a far more affluent country to ease the Russian and Eastern European transition from Communism in the years after 1989.)

That this decisive initiative was not called the Truman Plan is an extraordinary misjudgment of history. It was first proposed in a casual manner by George Marshall, the Secretary of State, in a few lines of a commencement address promising help and good will to Europe. Thus the Marshall Plan. In fact, it was Harry Truman who had responsibility for giving the project substance and for winning congressional support. It was Truman who would have been blamed had the Congress refused funds or if there had been anything less than success in the execution of the plan.

On other matters Truman brilliantly assessed risk and then acted. The most spectacular instance was the firing of General Douglas MacArthur in 1951 for insubordination and unwillingness to conduct a limited war.

For my generation MacArthur was one of the most celebrated figures, military or civilian, of the time. He had had a noteworthy military career in World War I; then, back in Washington, he led the attack on the Bonus Army, the encampment of veterans and their families seeking greater recompense for their wartime service. He went on to the Philippines as local force commander and in World War II to the Pacific Army command.

A less impressive man would have suffered from the

negative publicity that surrounded the Bonus Army action and from his failure to foresee and prepare for the near-certainty of a Japanese attack on the Philippines immediately after Pearl Harbor—a major display of military incompetence. He nonetheless survived, moving on from Guadalcanal north, island by island, to the Japanese surrender and eventually to his imperial years in Tokyo. With Eisenhower and Marshall, he was one of the three most illustrious military figures of the Second World War. After the outbreak of hostilities in Korea in 1950, he brilliantly envisioned the encirclement of the North Korean forces by the Inchon landings. Then came the disastrous rush to the Yalu and his devastating defeat by the Chinese. From success he had made his way to disaster for the forces he led and for himself. He reacted by making his truly alarming call for nuclear action against China, for laying down a radioactive defensive belt of undetermined nature along the northern border of Korea, and for intervention against China from Formosa by the militarily ineffective forces of Chiang Kai-shek. Calculating the truly global risks of keeping MacArthur in his post and the domestic reaction should he be removed, Truman made the correct assessment. He sacked MacArthur. The country as a whole responded favorably. MacArthur returned to address the Congress in his own defense and pass into an anonymity he much regretted.[2]

2. My only contact with him was in 1956, when, with other members of Adlai Stevenson's presidential campaign staff, I got word that if a public meeting with Stevenson could be arranged with adequate publicity, MacArthur would endorse him against Eisenhower. He greatly disliked Ike. We did not think it worthwhile to respond.

From a confrontation most Presidents would have avoided or evaded, Truman emerged triumphant.

The elimination of Douglas MacArthur, I might note, was not received with uniform approval. In 1964 or thereabouts, I was in Kansas City and went out to Independence to visit the Truman Library. I was marginally concerned at the time with the prospective library for John F. Kennedy and wished to see a distinguished predecessor. At the end of a long row of boxes filled with correspondence I took one down to see what it contained. It was letters protesting the firing of Douglas MacArthur. I put it back, strolled along to the other end of the row and sampled another box: inside were still more letters condemning Truman for dismissing MacArthur.

The Truman political judgment was not infallible. In the summer of 1952, he should have accepted the peace terms offered in Korea: the division of the peninsula on what were then the battle lines and what is still, effectively, the border between the two parts of the country. These, wisely, Eisenhower did accept a year later. A conservative can too often, as in the case of Nixon's recognition of Red China, take action that a liberal, in fear of conservative outcry, avoids. It was this fear that deterred Harry Truman. Equally adverse was Truman's response to the McCarthy-Nixon crusade against alleged subversives in government—the Truman loyalty program. This involved a search affecting all civilian employees in all government departments for those with disloyal tendency or background. I offer another small piece of personal history.

One day in 1950, the Truman years, I went from Harvard to Washington for a meeting at the Department of Commerce; it was held to discuss the effect on the economy of agricultural subsidy and price-support programs, a politically innocuous enterprise if ever there was one. Filing for travel expenses and compensation for a day of mild scholarly debate, I unknowingly made myself an employee of the Department, a consultant. Accordingly, I became the subject of a loyalty investigation. Of this I was also unaware until some months later, when I received a peremptory demand for word, perhaps a confession, as to my association with three suspect individuals. One of them I did not know; one was an old friend with impeccable credentials. The decisive black mark on my record was the third, Corliss Lamont, once a neighbor in New York City and long a very good friend. He was widely known as a radical but not for any Communist affiliation. I wrote to ask the reason for this request, saying that I had not known I was employed by the government. This was taken to be a resignation. In later times I petitioned for, and received, my FBI file; I discovered that my non-employment at the Department of Commerce had precipitated many weeks of investigation and had covered territory from Boston to New York to Washington and California. Friends, conscious of what was needed, described me as conservative; one said I was a reactionary. When I was later called to public service, including as ambassador, it was noted that I had resigned while under investigation.

The Truman loyalty program was no slight aberration. It ruined the lives of good and devoted civil servants. No sensible government accords insult to, spreads fear and despair among, those who do its diverse, difficult and important tasks. There were also specific consequences.

In earlier years a distinguished group of professional Foreign Service officers—John Carter Vincent (my close personal friend), John Stewart Service, John Paton Davies, O. Edmund Chubb—had seen the corruption and associated incompetence of the Chiang Kai-shek regime in China and told of the inevitability of a Communist takeover there. For this they were later fired or consigned to positions of no influence. When it was shown that they had been right, there was surprise but no forgiveness. Instead, in one of the more depressing exercises in the history of political invention, they were held to be the men who had lost China.

And there was a deeper, less visible effect of the Truman loyalty program. Seeing its consequences for certain individuals and fearing its intrusion on their own lives, many in the government sought protection by strongly asserting their anti-Communism. In the public action that ensued, policy was based not on reality but, instinctively or deliberately, on personal caution. This tendency survived and remained important in the foreign-policy attitudes on Vietnam, elsewhere in Indo-China, in the Middle East and in Latin America. Those who urged a militant and sometimes military anti-Communism were considered sound, trustworthy and personally safe; those who questioned such a course were politically unsafe, possibly even slightly disloyal. This was the least appealing part of the Truman legacy.

In recent years, Truman's decision to drop the atomic bombs has been the most debated of all his actions. I am not wholly in sympathy with much of this discussion. At the time and in the context, Harry Truman did not have a choice.

I first heard of this weapon in the summer of 1945, when, as I elsewhere tell, I was in Germany assessing the effects of the strategic air war. George Ball, back from a high-level meeting in Washington, told me we had just tested a new and devastating atomic device. (Through a slight error in his Western geography, he said the test had taken place in North Dakota.) Surveying the cruel and terrible wreckage of the German cities, with bodies decaying beneath the rubble, I had a brief hope that it still might not work. I had no thought that if it worked, it would not be used.

This was in keeping with the controlling mood of the time: weapons were developed to be used, and if they worked, they *were* used; any failure to do so was unthinkable. The nuclear bombs might be unimaginably cruel and possibly unnecessary, but so were the end-of-the-war raids on Dresden and the bombs and fire bombs on Tokyo, the latter bringing death and destruction greater than that which occurred in Hiroshima or Nagasaki.

We were caught in a wartime exigency that was beyond any humane, any rational, resistance. We now know that there was discussion of the use of the bombs in Washington that was led by Secretary of War Henry L. Stimson. The outcome, however, was predetermined and inevitable. Truman, had he resisted, would have encountered a force far beyond presidential power. It was as wrong, evil, to use the atomic bombs on innocent men, women and children, as, to repeat, had been the earlier saturation bombing. But war captures all thought, all restraint, all humane commitment. It did not spare Harry Truman.

After Harry Truman came Dwight D. Eisenhower. Seemingly, there could have been no greater change, or so it was thought at the time. In fact, the two Presidents complemented each other in admirable fashion. Both were men of political experience—experience in the art of assessing risk and winning support for desired action. Truman's ability in this regard has been sufficiently noted. Although seen as someone who had stood politically aloof, Eisenhower had also been wonderfully indoctrinated in the less than merciful politics of high military command, specifically that of the Allied Commander-in-Chief in Europe. There he had learned to handle far-from-compliant military leaders and to some extent the dominant political figures of the time. No one who had close association with Churchill and Roosevelt, Bernard Montgomery and George Patton or, for that matter, George Marshall lacked compelling instruction in the art of politics.

With Eisenhower and the passage of time, the Truman loyalty program was watered down. After a final explosion, the anti-Communist crusade of Joe McCarthy came to an end, along with Joe himself. Peace came to Korea on the terms that Truman had rejected. And, a major and already-mentioned point, under Dwight D. Eisenhower the social programs of the New Deal—Social Security, unemployment compensation, public housing, the diverse welfare initiatives—were fully accepted.

As was the case with his reference to the dangers of an emerging military-industrial complex, Eisenhower could be admirably clear. On other matters it was the Eisenhower genius so to state his position that no one could be sure what he intended. Milton Eisenhower, his brother and a friend of mine, once told me that Ike's indirection in speech was a deliberate design for keeping open his course of action or for confusing the opposition. He remembered a meeting in the Oval Office at which some difficult and potentially very unpopular decision was reached. Reflecting on the expected adverse reaction, Ike had said, "It's all right. When I've explained it to the press, no one will have any clear idea what we intend to do."

Dwight D. Eisenhower was, in many ways, the right man for his time—as noted, the Republican who made permanent the New Deal and the emergent welfare state. I did not then so see him; within the limits imposed by talent and occupation, I worked harder in opposition to him over the years than against any other political candidate, with the exception of Richard Nixon. Much of this effort centered on Adlai Ewing Stevenson, to whom I now come. He was one of the most interesting political figures in the second half of the century. Greater attention should, however, be given to his principal misfortune, which was that he was twice pitted against one of the most effective politicians of the time, a conservative who accepted, confirmed and carried forward the liberal agenda of Franklin D. Roosevelt and Harry Truman.

7

Too Madly for Adlai

To this day Adlai Stevenson, twice a candidate for President, twice soundly defeated, continues to be actively discussed by older members of the American polity. No modern politician, not even Roosevelt or Kennedy, had a more faithful coterie of supporters. And Adlai was not only supported; he was loved. When he lost to Eisenhower in 1952, there was no doubt in the minds of those around him that he would run again in 1956; his immediate constituency even continued to rally to his candidacy four years later, the two earlier defeats notwithstanding.

Adlai Stevenson, more than anyone I have known, was a truly engaging person. His greeting was warm, affectionate; he really wanted to see you. Conversation with him was easy, giving always the impression that he intended to draw on your knowledge and intelligence. He had an evident concern for both the local and the world scene, on which he let you believe you were usefully informed. He had a quiet but wholly responsive sense of humor. Adlai spoke of his recent and current reading, with a special enthusiasm for the books he was about to read. No one ever visited Adlai Stevenson or heard him in an intimate gathering without a feeling of pleasure—and rarely without telling of this to the less favored.

Sadly, he did not reach out as well to the public at large. As already observed, Harry Truman was of the people, while Adlai Stevenson was apart and above. This, to a marked extent, was his own view of himself. He looked with mild disapproval at politicians like Truman and, later, the Kennedys who, in manner, speech and program, sought identification with, acceptance by, the masses. That was not for him. He was a leader; a leader stood above the crudities of the political scene. He had shown in the morass of Illinois politics that he could do so and still be elected; it was not significant to him that there he had, in some measure, been selected by the Chicago machine. He was sure that an informed leader was what the country as a whole deserved and wanted. His appeal was very powerful indeed to a privileged elite; it was not as strong, alas, to the greater voting public.

Nor did he make concessions on substantive issues in order to achieve and sustain a wider popularity. For much of his adult life he had lived in suburban Libertyville, Illinois, with affluent Chicago neighbors. He was well regarded by them, and this regard he returned, at least in part. Accordingly, it was with caution, even reluctance, that he accepted the economic and social legacy of the New Deal and the evolving welfare state. As governor, he had been a glowing alternative to the

An amused walk with L.B.J. from Oval Office to White House.

The 1960 Democratic Convention. I was informing J.F.K. for a meeting on agriculture, a matter of which he did not wish to know.

With Robert Kennedy at a press conference in a military hospital in Hawaii. I was suffering a short-term Indian affliction.

Meeting with Nehru and David Bell, the head of the Agency for International Development. Also visible, Mahatma Gandhi. The discussion was of economic aid.

Nehru again. A seemingly intense exchange, probably during the Indochinese war in 1962.

Jackie views a cheap tin elephant. She wrote on the photo, "You promised to show me the Treasures of India."

A rare meeting with Harry Truman. Fellow liberals crowd in.

Boarding a plane in New York with L.B.J. and Lady Bird. J. Edgar Hoover is deeply disapproving.

Eleanor and friends.

With Gene McCarthy at the wild 1968 Democratic Convention.
I'm about to second his nomination. (Burton Berinsky/*Time*)

Temporary Building "D"

January 29, 1942

Refer to: 5:12:RMN

RECEIVED
JAN 31
FILES
OPA

Wilde's Standard Service
Oregon
Illinois

Gentlemen:

In reference to your letter of January 20, an amended order, effective January 28, allows a dealer in tires and tubes to transfer his stock to a wholesaler or distributor provided that he keeps a record of the transfer showing the name of the person acquiring the inventory, the number and type of tires and tubes acquired, sales price and the date of transfer.

Consequently, the transfer to your distributor in this case does not require special permission of this Office.

Very truly yours,

J. K. Galbraith
Assistant Administrator

SENT
JAN 30 1942
FROM PRICE DIVISION OF
OFFICE OF PRICE ADMINISTRATION

By Ben W. Lewis
Price Executive
Rubber and Rubber Products Section

RMNixon:ms

WRITTEN BY ....
REVIEWED BY ....
REVIEWED BY ....
REVIEWED BY ....
REVIEWED BY ....

1942. Note carefully the lower left-hand corner. Richard Nixon was writing my letters on wartime rubber rationing, of which I was then in charge.

sordid, graft-addicted politicians of Chicago. This, not a firm embrace of the New Deal, was his claim to the presidency. Although his political restraint was recognized, it did not diminish the respect and affection that socially committed liberals accorded him.

The most evident manifestation of Stevenson's detachment from the politics of the masses, or his desire to seem detached, was his well-publicized reluctance to get into the presidential race at all. In the weeks and months before his nomination in 1952, he made an impressive showing of his doubts. These he carried fully to the press and the public. Also to President Harry Truman when Truman, to his later regret, urged him to run. How better prove one's respect for the presidency than to avoid any display of personal ambition? That office should seek the man. I've always supposed, as have others, that he gave considerable thought to making this point.

While Adlai Stevenson rejected the mores and manners of the conventional politician, he exploited fully his own different appeal. His most frequent and to many his most endearing political statement, often repeated, was "I want to talk sense to the American people." Holding them to be intelligent, he held that they were to be informed, not persuaded, and certainly not by mindless rhetoric. Being intelligent and knowledgeable about the issues, they would listen carefully, come to the right conclusions and vote for him. Again, the reaction was different as between Stevenson's immediate following and the public at large. The former rejoiced in the idea of responding to straightforward relevant information and the appropriate action without any of the customary political rhetoric. Rhetoric appeals to a wider, less reflective audience.

All this shaped the form and content of the Stevenson campaigns. With them I was closely involved, more so than in the case of any other candidate, John F. Kennedy not excepted. The error of our course I came only gradually (and reluctantly) to see.

After the 1952 campaign Stevenson's speeches, much admired, were published in book form. Stevenson sent me a copy; I was well represented therein, and I put it on a shelf with my other writing. Looking at my contribution in later years, I sensed the fatal flaw: I had written seeking the approval of the candidate, his campaign acolytes and myself, not that of the larger public, the electorate as a whole. So had the other writers.

In July 1952, I was in California, attending a conference on rural development problems. Such conferences are a way of life in the academic world; a latter-day Veblenite called them "The Leisure of the Theory Class." Thus I was safely away from the Democratic Convention, which was held in Chicago that year; I was still cultivating my commitment to solemn academic pursuits and had made no effort to be a delegate. However, my scholarly dedication was seriously eroded when I received two telephone calls, one from Arthur Schlesinger, Jr., the other from George Ball, both of them headed for major distinction in American public life. Each was unaware that the other was calling. Each told me that Stevenson's nomination was now secure, and I must come immediately to Springfield, Illinois, to join the candidate's staff and speech-writing team. I had never met Stevenson, although I had followed his political career with interest and approval.

Returning from California by train, I disembarked at a remote stop in northern Illinois, took a bus to Springfield —campaigns were then conducted at low cost—and went to my hotel. There I was told by Schlesinger, to my considerable distress, to remain in my room until further notice. I began the Stevenson campaign with house arrest.

At a press conference the day before, making note of Schlesinger and others who had come to help, reporters had asked Stevenson if they were "taking over his campaign." And why, a very delicate point, did a man of such oratorical talent need speech writers? Could he not rely on himself? Stevenson had been deeply upset; he put a hold on the hiring of more personal staff. That, in particular, meant me.

Toward evening word came that I was to be released. The candidate had been persuaded, not without difficulty, that, given the pressure of time and effort, he would, indeed, need literary help. For all public purpose we would be designated research workers; no one could seriously question the need for research. I was taken to a garden party and there met Adlai Stevenson for the first time.

As were so many others, I was immediately impressed. Then slender, good-looking if not handsome, he quickly made me at ease in his company. As earlier observed, one had a sense of prompt, informal, intelligent access to his views. This would continue to be the case in the next thirteen years of our friendship.

During the campaign I wrote Stevenson's speeches on economic matters and, as a sometime agricultural economist, those on agriculture, which was then an important matter. Stevenson, though governor of a leading farm state, was, like most urban politicians, sadly innocent on farm issues. On occasion, I traveled on the candidate's plane, and when Harvard resumed classes in the autumn, I continued to commute to Springfield.

Once, as the weeks passed, I had a vision of success. Catherine Galbraith was taking me to Logan Airport in Boston for a flight to join Stevenson in Illinois for some seemingly urgent effort, and we ran into a traffic jam at the Sumner Tunnel entrance. There was total deadlock, not relieved by the constant din of automobile horns. Eventually I got out and told a policeman of my problem: I was a vital force for Stevenson and the Democratic Party; I must get to the plane. To my amazement, he took command with both authority and skill and, incredibly, cleared a path into the tunnel. He saluted handsomely as we were on our way, and soon I was safely aboard. I assumed that this meant we were getting through to that large constituency to which Roosevelt and Truman had so appealed. Alas, it wasn't so.

One important difficulty did become evident to me, as to others, in the early days of Stevenson's campaign. It was still for us the high noon of the New Deal; for Adlai it was not. His enthusiasm for extending and perfecting Social Security, the greatest of the Roosevelt innovations, and the welfare state generally was decidedly muted. The labor movement, then at the zenith of its power, wanted the repeal of the Taft-Hartley Act, which had been passed in 1947 by a Republican Congress over a veto by President Truman. It banned jurisdictional strikes, secondary boycotts, the closed shop, and imposed other restraints on the unions. Repeal was high Democratic gospel; Stevenson was adamantly opposed. After hours of persuasion, he would say only that it should be replaced, although it was never clear by what. Once he suggested that, if necessary, he would forgo endorsement by the unions. It fell to me and rather more to Willard Wirtz, a devoted supporter with close ties to labor, to persuade him that the backing of the unions was both acceptable and deeply essential.

There was also a more general economic problem. In his college years Stevenson had studied at Princeton under some of the country's most dedicated defenders of eighteenth- and nineteenth-century classical economics. John Maynard Keynes and the New Deal had now become part of the liberal orthodoxy, and this counseled that in a depression, which was a prevailing fear, you might unbalance the budget in order to put people back to work. Stevenson was appalled that anyone should advocate such a reckless attack on reputable thought.

Again it was my job to urge the modern case. A return to old, discredited beliefs of earlier times could not be accepted; we must be prepared for another and prolonged depression and have a policy to alleviate it. I recall my effort at persuasion without pleasure. After we were defeated in 1952, as I will later tell, a major attempt was made to bring Stevenson's economics into the twentieth century as we saw it. During his first campaign, Stevenson remained with the past.

On the international scene, there was the more urgent issue of the Korean War, and here the conventional liberal view had a truly damaging effect. Had it not been for the subsequent darker cloud over Vietnam, the Korean War in its later phase would now be celebrated as the most unpopular conflict in American military history. By 1952, after the defeat and discharge of MacArthur, the war had settled into a military deadlock along what was to become the more or less permanent boundary between North and South Korea. In the presidential campaign Eisenhower had promised, if elected, to go himself to Korea and seek a settlement. This, as an undoubted war hero, he could do without being accused of appeasement. Stevenson had responded by voicing the accepted doctrine: the general should go instead to Moscow, for that was the controlling power in the Communist world. There, any decision concerning the war would be made. Ike's promise was quite possibly decisive in his winning the election.

Present here also was the already-mentioned advantage of conservatives in these matters. They could act as liberals could not. In later years, in withdrawing in defeat from Vietnam (however slowly) and in opening relations with Red China, conservatives, exempt from the thought and accusation of being soft on Communism, would take action that liberals felt forced to avoid.

Such was Stevenson's situation in 1952. Under attack from Joe McCarthy, who thoughtfully referred to him as Alger Stevenson, a reminder of his rather casual support of Alger Hiss, Adlai was forced to give his reluctant support to the war in Korea. Here, to repeat, was Eisenhower's strength against Stevenson, as it would be still in 1956. Blessed are the peacemakers; they also inherit the votes.

In American politics those who lose elections, presidential elections in particular, disappear from the larger scene—Michael Dukakis, Walter Mondale, George Bush, even in some measure Jimmy Carter. In this respect, Adlai Stevenson was all but unique. After the 1952 loss and his graceful concession speech, paraphrasing Lincoln — "[I'm] too old to cry, but it hurts too much to laugh"—the accepted task of his supporters was to prepare for the next presidential contest. We assumed that our respect and affection would eventually be shared by the larger voting public. Belief, as so often happens, was pleasantly disassociated from thought.

We recognized, however, that there was work to be done. Stevenson had to be separated once and for all from that early education in economics and from the moderate and agreeable conservatism of his Libertyville neighbors. The result was one of the most engaging exercises in modern politics and certainly the most remarkable in which I have ever participated. It was a tutorial seminar designed to teach Adlai the fundamentals of contemporary and, as we believed, politically controlling liberal thought.

Called the Finletter Group for Thomas K. Finletter, New York lawyer and former Secretary of the Air Force who served as host and convenor and the source of the modest amounts of money required to support the project, it was made up of a dozen or so scholars and scholarly politicians, most of whom, as the one in charge of such matters, I recruited. We met at Finletter's Manhattan apartment with (and often without) Stevenson. When he was there, he was instructed on the accepted economic and social thought of the time and the consequent political action; when he was not, papers were prepared, discussed and sent to him to be studied. As already told, Stevenson had endeared himself to intellectuals with his praise of books and articles he was just about to read, and this was the case with our memoranda. However, he could not so escape when he was present at our meetings. On one memorable occasion, Professor Paul Samuelson of M.I.T., the most respected economic figure of the day, who, with Alvin Hansen of Harvard, had brought John Maynard Keynes to American economics, lectured him firmly and at length, as he might any eager student. Reviewed and stressed were the basic concepts of modern economic thinking. Stevenson seemed impressed. So, perhaps optimistically, were we.

Our confidence was slightly shaken in the months immediately before the Democrats were to convene in 1956. Estes Kefauver, senator from Tennessee, a strong populist liberal, on occasion given unduly to alcohol, emerged as an opponent. He swept an early primary in Minnesota. Adlai was forced to shed the mantle of reluctant savior and campaign vigorously for the nomination, particularly in California. By the opening day of the convention, the nomination was his for a second time. On an open vote, a remarkable concession to democracy, Kefauver narrowly defeated John F. Kennedy to become the vice-presidential candidate.

This time the campaign headquarters was established in Washington; I joined the staff, commuting back to Harvard for my classes. As before, I wrote the speeches on economics and agriculture and, as an added assignment, those taking an adverse view of Richard Nixon as the sitting Vice President and the vice-presidential candidate. The latter I did in response to Stevenson's truly heart-warming request, which on other occasions I have recalled: "Ken, I want you to write the speeches against Nixon. You don't have this tendency to be fair." The most notable of these Nixon rebukes was delivered on October 27, 1952, before a vast audience at the Gilmore Stadium in Los Angeles:

> Our nation stands at a fork in the political road. In one direction lies a land of slander and scare; the land of sly innuendo, the poison pen, the anonymous phone call and hustling, pushing, shoving; the land of smash and grab and anything to win.
>
> This is Nixonland. America is something different.

As I reread these lines after more than forty years, I find myself yielding to Stevenson's initial judgment. Seeing them just before they were to be delivered, he said, "This is the kind of speech that can only lose us votes." Then, noting the hour, the lack of any time for changes, he cleared it with a fine Stevensonian comment: "I suppose we might as well tell the truth about the man." The speech, judging only from the response of the huge audience, was successful. But almost certainly it did not change votes; it had been given to an audience of the already committed. That, as we were coming to know, was the losing side.

Dwight D. Eisenhower was still considered the country's best hope for continuing peace, as he had been four years earlier. It is true that John Foster Dulles, his Secretary of State, was not similarly so celebrated. His violent anti-Communist rhetoric, his network of defense alliances, his militarization of the Third World, were all repellent. But here, again, it was not for liberals, much as they might privately be critical, to lead the attack. So, Dulles notwithstanding, Ike was seen as a bulwark of peace.

Once during the 1956 campaign I remained behind

after a major farm speech by Stevenson in Iowa. It was at a plowing match, a ceremonial occasion of no particular partisan character. I had written the address and I wanted to sense the reaction; the Eisenhower Administration had been thought weak on its farm program. As I questioned one listener after another, the dominant response, four years after the end of the war, was that Ike had got us out of Korea. For a farm audience at a farm speech in the heart of the farm belt, farm issues were less important.

In 1952, after twenty years of Democratic Presidents, we had assumed as a matter of course that we would win; losing, I had suffered a depression. By the end of the 1956 campaign there was no such expectation of success. This time in the last days I had a bad intestinal upset; on election night I was in a Boston hospital. I didn't turn on the radio to hear the results.

Adlai was in better spirits in those final days, and this is a good illustration of something regularly present in the political process. Nothing so protects a losing candidate from knowledge of his probable fate as the mechanics of the campaign. He is surrounded by an admiring staff and highly articulate advisers. They are wonderfully optimistic; not theirs to convey the bad news even in the unlikely event of their considering it. The candidate comes into town through enthusiastic, cheering crowds led by the high school band. Who can have doubts amidst such a display of popular approval?

And there is a further question. Who, while experiencing such acclaim, can have doubts about the pleasures of political life? It is clearly better to win, but even a losing campaign is a compelling experience. It is far, far better than not running at all.

After the 1956 defeat Adlai Stevenson retreated from politics. He joined a New York law firm and called me once for help on a legal case—an antitrust prosecution, as I recall. I saw him on social occasions, but in the next years I moved to the support of John F. Kennedy. A loyal coterie, not a few of its members women, remained with Stevenson, believing, as I've already indicated, that he should be the candidate for a third run. Affection, love, overrode political judgment. I was regarded and, indeed, assailed as a morally defective defector.

At the 1960 convention, which nominated John F. Kennedy, a small Stevenson headquarters was staffed and led by George Ball, his greatest source of guidance and support in the earlier campaigns. Were there to be a convention deadlock, perhaps Stevenson would sweep the floor. Adlai himself did little to further the possibility. A speech he made to the convention was trivial in the extreme, principally citing the intense crowding around the several campaign headquarters at the Biltmore Hotel. However, the Stevenson personality and its appeal survived. He continued to have his devoted circle of friends and admirers. This would be a decisive factor in his relations with John F. Kennedy.

*

When dealing with the political issues of the time, Adlai Stevenson was most at home, most confident, in foreign policy. Those of us who wrote the speeches on domestic matters had a range of discretion not allowed to those writing on foreign affairs. Accordingly, he seemed the obvious person to be Secretary of State in the Kennedy Administration. This, we all assumed. Not, however, J.F.K.

We did not fully realize that Kennedy and Stevenson were never at ease with each other; there was a subtle clash of personality. More important, if Adlai were Secretary of State, Kennedy felt that he would continue to appeal to his own separate constituency, the well-recognized Stevenson circle. Better someone of lesser stature who would be fully at the Kennedy command—or so it was hoped. There was also the familiar tendency in politics to think better of those you do not know; you have heard much of their virtues, less of their faults. So the Kennedy choice became Dean Rusk, a man solidly in the Cold War tradition, a faithful supporter of the Third World military alliances and later, alas, of the Vietnam war.

Still, Stevenson and his coterie could not be ignored by the Kennedy Administration. The solution was to raise to Cabinet stature the post of Ambassador to the United Nations and give it to Adlai. This, with no great pleasure, he accepted. Within a few weeks he was ordered to make a palpably mendacious defense of the Bay of Pigs adventure, which was revealed as such within a few hours. After the Missile Crisis was over, in 1962, he was brutally assailed for having suggested that if the Soviets would remove their missiles from Cuba, we would remove ours along the Soviet border in Turkey. This was held to be appeasement of the most feckless sort; Adlai's Munich, it was called. Later, it became known that the suggestion was a key part of the bargain Kennedy had made with the Soviets, but he had thought it politically wiser not to have it divulged at the time.

In New York, Stevenson's friends were faithful and attentive. His social life was expansive; so became his figure. He survived Kennedy by a little less than two years, to fall dead on a London street. Two days later there was a service of remembrance and farewell at the National Cathedral in Washington and never a deeper manifestation of sorrow. In political life many are respected and more are cultivated. Not so many inspire love.

8

John F. Kennedy

I HAVE ALREADY sufficiently observed that Adlai Stevenson's political personality, while it attracted a committed corps of supporters, did not capture the country as a whole, not even all the traditional Democrats. No one saw this more clearly than John Fitzgerald Kennedy. He too was from a highly privileged background—a father with far more money than anything known to the Stevenson family, private preparatory schooling, Harvard University, a visibly affluent way of life. This was not the usual aspect of a man of the people. Of that, Kennedy was also deeply aware.

But offsetting this aura of privilege, and contributing to the Kennedy political success, were some solid advantages. He was young and handsome; he had a beautiful (and singularly intelligent) wife; he was a war hero. More important, he had been educated in the demanding culture of Massachusetts Irish politics, and by the time he ran for President, he had had considerable experience in the political behavior patterns of the House of Representatives and the United States Senate. These he had studied with detachment, humor and, more than occasionally, privately expressed contempt. He was also intelligently abreast of the major issues that a presidential candidate would encounter. Over the years, more than Stevenson, he had learned where to turn when in need of information or guidance. In one of the more dreary characterizations of our time, he was a quick—a very quick—study.

Kennedy also knew how to identify himself with an audience and with the larger electorate. At the end of his 1960 campaign, he addressed a vast crowd in the old Boston Garden. As a member of his campaign staff, I was there. He asked himself, as though from the floor, why he was running for President. In reply, he listed some issues, all relevant to his audience, that needed attention; then he ended by saying that the presidency was a well-paid job with no heavy lifting. The largely working-class gathering responded with appreciation, affection and joy. He was one of them.

There was another basis of John Kennedy's political strength: he was intensely, even uniquely, true to himself. Of the importance of this he was acutely conscious. In the summer of 1960, soon after the convention, Arthur Schlesinger, Jr., and I traveled to Cape Cod to discuss with him the coming campaign. Without doubt, a very demanding time lay ahead. Reflecting on this, Kennedy said he would have one great advantage over his opponent: "Nixon has always the problem as to who he is. I know who I am."

The pages that follow, a look at the compelling problem of Vietnam excepted, are on Kennedy as a person and politician and not on the economic and political issues of his time. On these there has been ample comment, some from me.

All politicians surrender some personal belief to political necessity. The common error, then, is that the needed concession is turned into belief. What is politically advantageous becomes the truth.

Not so with Kennedy. He made political concessions, but he retained his own convictions. These he was regularly led to proclaim to immediate and trusted friends, sometimes to their surprise, even shock. This candor he extended to the matter of his religious faith.

It is regularly assumed by non-Catholics that the Church and its hierarchy invite a solid loyalty; that is what makes a good Catholic. Kennedy was more selective. He and members of the Kennedy family were strong in their support and affection for Richard Cardinal Cushing of Boston. They were not at all on good terms with Francis Joseph Cardinal Spellman of New York, a devout conservative. His disapproval they quietly accepted as good Catholics should but also as politics required.

At Kennedy's inauguration, Cardinal Cushing seized upon the occasion and the large audience for a very lengthy prayer. A day or so later I toured the White House with the new President—no longer Jack but the President. He was looking at the real estate, as he called it, that he had just acquired. He asked me how I had liked "the Cardinal's speech to God." I responded in a noncommittal way; I had entered the world as a Covenanted (Hard-Shell) Baptist, a faith that then had a strong anti-Catholic bias. I had been instructed by my father, who was politically engaged—as I will later tell, a political boss—to keep what I heard in church to myself. Kennedy was more relaxed. "I knew it would be long," he said, "but halfway through I was saved by the thought that here is Kennedy, the first Catholic President, being inaugurated, and Cardinal Spellman is having to watch it on television."

My mind returns also to a White House press briefing prior to one of the first presidential press conferences. The main subject, predictably, was foreign policy; in those days it always was. The State Department had sent over the recommended responses to the probable questions. Kennedy looked at them and reacted with genuine anger. "I would like a few answers," he said. "I don't need any instruction on how to evade."

While willing to dissemble as necessary, he preferred the truth. I especially remember a conversation following the Cuban Missile Crisis, when I was back briefly in Washington from India. We had gone to the theater together, where we retreated to the backstage steps between the acts to escape the well-wishers and the autograph seekers: "Mr. President, I didn't vote for you, but I sure admire your stand on quite a few issues."

We talked of the crisis just past and of those—the generals and the Cold War liberals—who had wanted to bomb the missile sites and possibly more, thereby accepting the risk of a nuclear war. "You have no idea," he said, "how much bad advice I received in those days." There was later to be similar forthrightness when he discussed the political, military and staff guidance to which he was subjected on Vietnam.

Another example of Kennedy's candid thought and speech came to me from Carl Kaysen. Long a distinguished professor at M.I.T., in the Kennedy years he was an influential White House assistant on varied assignments, including the ongoing and politically charged negotiations on the Panama Canal. On this issue Roberto Francisco Chiari of Panama had come to Washington for talks with Kennedy in June 1962. The two Presidents met for a time in the early morning; Kennedy had then to go to another engagement. Before their next meeting, at noon, J.F.K. asked Kaysen for some additional information he needed, commenting that the talks weren't going well. Kaysen expressed concern and asked what was wrong. Kennedy replied, "He says we've been screwing them all these years, and I agree."

Kennedy's preference for plain talk did not spare his friends. Before I left for New Delhi, in April 1961, we had a farewell breakfast at the White House. That morning the *New York Times* had a piece on the new Ambassador to India; Kennedy asked how I liked it. It had been generally favorable, and I said it was all right, but I didn't see why they had to call me arrogant.

"I don't know why not," said Kennedy. "Everyone else does."

I first came to know John F. Kennedy in the late 1930s, when he was a student and I a young resident tutor at Winthrop House, one of the several Harvard residences. Admission to the houses at that time was by application, and Winthrop House was greatly favored by athletes, then as now called jocks. It was also a resort of undergraduate students of Irish paternity, a welcoming haven from the prevalent social attitudes of the time. In later years in an address at Harvard commencement—the Class Day speech—I reviewed my life at the university, the improving social mood, and made the unfortunately valid observation that Winthrop House had prided itself in my time on being the first of the Harvard houses to go on from being anti-Irish to being anti-Semitic. I was sharply criticized; even at a university some historical truths must be left unsaid.

All four of the Kennedy brothers came eventually to Winthrop House; Joseph Kennedy, Jr., the oldest, who was later killed on a dangerous experimental air mission during World War II, was one of the closest friends I ever had among Harvard undergraduates. As such, he was my opening to the Kennedy family and to his brother John Fitzgerald. The latter, a couple of years younger than Joe, I knew much less well at the time. The younger Kennedy was not strongly committed to academic work. Nor was he much interested in the political scene, which in those years of the Great Depression and the Roosevelt salvation was very much a part of our lives. He had, instead, his friends and his social life, which commanded most of his time. Of this, his brother was more than occasionally critical. So was his father.

One of the sharpest references to Jack Kennedy I heard in those days was from Joe, Jr. He came into my room one evening in the autumn of 1936 to tell me he had in mind a sure way of making money. He invited me to join him in the enterprise; I declined but remained in touch.

Joe intended to exploit financially a deep human failing; that is, the tendency for hope to control expectation and good sense. Of this, there was a promising current example; those who hoped that Alfred Landon would win the election against Roosevelt in the next weeks were allowing their hope to color their expectations. By taking bets for Roosevelt and against Landon, one had a sure thing. Joe set up shop in the Harvard Business School, a favorable territory, and after the votes had been counted, he had made enough money to buy a new car. To celebrate, he invited me to ride with him to visit some friends in Wellesley. He was in a sour mood as regards his younger brother. Jack had also wanted a car and had bought one on, as it was then called, the installment plan.

Joe said, "Here I work hard to get a new car, and he gets one on credit."

*

The war came, and I was carried away to Washington and the wartime bureaucracy. The Kennedy brothers disappeared from my view; their father was now in official disrepute because, while Ambassador to England, he had written off the British in what was deemed an irresponsible and prejudiced way, as indeed it was. My association with the Kennedys was not renewed until 1946, when J.F.K. emerged as a congressional candidate for Cambridge and the Massachusetts district of which it is a part.

I did not participate actively in his early campaigns, either for the House of Representatives or, later, for the Senate. He was a somewhat remote figure, relying on his family, its money and a growing corps of aides whom I did not know, including the members of what would be called the Irish Mafia—Lawrence O'Brien, Kenneth O'Donnell, Dave Powers. (In sharp ethnic contrast, there was also Theodore Sorensen, who would be a source of legislative and political guidance, good speech writing and a certain quiet discipline.) Finally, there was his younger brother Robert Kennedy, by far the most influential figure in the Kennedy political entourage. All these were later to become the inner circle at the White House.

My principal contribution, as I became involved, was as a source of information on economic issues. In J.F.K.'s congressional years, our relationship passed through three stages, wonderfully illustrating his emerging competence and mastery: at first he would call to ask how he should vote on one or another piece of legislation. Later he would call and ask for relevant background information—what he should read, to whom he should talk for guidance. At a final stage he rarely called at all; he had come increasingly to rely on his own knowledge and judgment.

In the late 1950s, our association was renewed as he became involved in his presidential campaign. For its January 1960 issue, *Esquire* magazine, then enjoying some modest political influence, asked fifty-four men and women of presumed literary and scholarly prominence for their preference among the presidential candidates to succeed Dwight D. Eisenhower. Adlai Stevenson led the field by a large margin—a third campaign. Following were Richard Nixon and Hubert Humphrey. John F. Kennedy was a poor fourth, with only five votes. Mine was one. Another was that of Crane Brinton, the eminent Harvard historian. He wanted a candidate who would not support "orthodox or 'classical' economics [but] what I call neo-Keynesian or Galbraithian economics." Kennedy, who took the poll seriously—more seriously than did I—was very grateful and welcomed me warmly as a supporter.

I had, it turned out, another needed quality. Because they were so obviously of the Irish community, Kennedy and the members of his inner circle were acutely conscious of the need for ethnic diversity. The campaign could not be too visibly an Irish endeavor. When I arrived in Los Angeles in July 1960 for the convention—as mentioned earlier, I was then co-chairman of the Democratic Advisory Council of the Democratic National Committee—I was immediately rushed to the Kennedy campaign headquarters at the Biltmore Hotel. There, I was supplied with floor credentials and put in charge of Kennedy delegates from northwest of the Mississippi. Here, I thought, was recognition of my political influence and skill. Then I was told by Kenneth O'Donnell, "You know why we needed you. We got out here and found we had no one on the whole convention staff who wasn't Catholic, Irish or Jewish." (Sorensen was drafting speeches and giving less visible guidance; the need to include an Afro-American was still some years away.) The following days were for me a powerful display of Kennedy's political energy and aptitude.

What was involved was perhaps the most diligent, even relentless, canvass of convention delegations ever made. Not one of them, committed or uncommitted, was left unattended. Its particular concerns were identified and, to the extent possible, discussed. At a meeting of the Kennedy staff each morning, the work of the day before was reviewed, that of the present day assigned. Once when I observed that we seemed to be comfortably ahead, I was sharply rebuked; the Kennedy operation did not allow for optimism.

Because of my background as an erstwhile agricultural economist, I was assigned one day to speak to a special caucus of delegates from the farm states. The subject was farm legislation, the farm program; it was one of the few matters on which Kennedy did not feel at home—"Where I was raised, you were taken out on a bus to see a cow." William Robert (Bob) Poage of Waco, Texas, a long-time member and then the chairman of the House Committee on Agriculture and a Lyndon Johnson delegate, offered a stemwinding oration on the importance of keeping the farm program under a statesman from the farm country, not an innocent urbanite from the East. I had prepared the case for Kennedy and was scheduled to speak next. I would urge that the best friend the American farmer had ever had was a liberal Easterner, Franklin D. Roosevelt. Before my turn arrived, however, Kennedy himself came in. He had decided that this was too important a matter to be delegated. He listened briefly to my proposed comment and then developed it, adding his own thoughts and leaving no one in doubt as to both his agricultural concern and his competence. Some in the gathering would still vote for others; Kennedy would henceforth be fully worthy of their trust.

The ultimate manifestation of the Kennedy astuteness during those July days was the selection of Lyndon Johnson as his running mate. The Kennedys didn't much like Johnson; Robert Kennedy had described him as a bore—a serious inaccuracy, for that precisely was what Johnson was not. His presence on the ticket would undoubtedly bring Texas to Kennedy and help elsewhere in the South. His influence in the Congress might also result in votes. The problem was the deep differences in the political style of the two men. But here was a constituency that Kennedy needed. So, against the advice of many around him, including his brother, Kennedy named Johnson the vice-presidential candidate.

I was less shocked than most when I was called early in the morning by Robert Kennedy, who told me in a distinctly angry voice that it would be Johnson. I would have to help calm the expected liberal reaction. My own reaction was not unfavorable. Johnson was of my generation; we had both been in Washington in the 1940s as committed New Dealers. By the time I reached the convention floor, I had my case in hand. It was Roosevelt again. "This is the same as in 1932, when F.D.R. picked John Nance Garner, also a Texan." If Roosevelt was right, why not Kennedy?

There was a negative liberal flurry, but it soon subsided; later Kennedy ticked off to me with his usual candor the reasons for his choice:

"First, I had the feeling that Lyndon takes more of the Catholic flavor off me than anyone else. A Southern Protestant.

"Second, he obviously helps me in the South.

"Third, it wouldn't be worthwhile being President if Lyndon were the Majority Leader."

The clearest expression of John Kennedy's political acumen was on the Vietnam war. I was opposed to our intervention there; so, without a doubt, was he. The difference was that he had to contend with the Pentagon warriors, some equally militant civilian advisers and the larger conservative political community. I did not.

My opposition was based on a closer and longer experience with that part of the world than that of most of my colleagues. I was not impressed by the threat of Communism in a country that had never had capitalism, democracy or even independence. From my knowledge of India and its problems (as I will later tell, I had had an advisory role there before I became Ambassador), I knew how powerful was the fear of a new imperialism in once-colonial countries. Such had, anciently and disastrously, been the history in Vietnam. We would be seen there as the successor to the Chinese, French and Japanese.

Back from India on a short visit in the autumn of 1961, I saw, somewhat illegally, the first, very secret proposal to send American troops to Vietnam. They were to go in under the guise of flood-control workers. I entered a vehement objection, and Kennedy sent me to Saigon to make a firsthand report; he did so knowing that I didn't have an open mind. I did not disappoint him. (My adverse comment eventually achieved publication in *The Pentagon Papers.*) In later months we were in frequent communication on Vietnam; he spoke of his desire to limit and eventually end our military commitment there. He told also of the civilian and military pressures to which he was subjected. What he heard from the Pentagon he thought in some ways predictable. He had less sympathy for the civilians who yearned to show that they could be as tough as the generals when it came to military decisions.

The debate as to what would have happened in Vietnam had Kennedy lived has continued. Speaking from a better vantage point than most, I have no question. He had no personal liking for this mammoth misadventure. And reflecting on the advice he had received on the Bay of Pigs and the Missile Crisis, he had a healthy mistrust of both civilian and military guidance. Just before his death he made clear his intention to bring our involvement in Vietnam to an end. It was no idle gesture. I do not doubt that he would have done it.

On November 22, 1963, Arthur Schlesinger, Jr., and I were invited to a meeting with Katharine Graham in New York for a discussion of the literary pages of *Newsweek*—the back of the book—with which, as the effective owner, she was not satisfied. Two or three editors were present. As we talked, the door opened, and a slightly diffident staff member asked if he might interrupt. There was news that Kennedy had been shot. Presently word came that he was dead. We flew back to Washington and went on to the White House. Later I went to Andrews Field to meet the plane, Jacqueline Kennedy and Lyndon Johnson returning from Texas. Then back to the White House.

The sorrow of those there gathered in the next days was, in a way, mitigated by the work now pressed upon us. The foreign visitors, the Capitol Hill ceremony and particularly the funeral commanded nearly total attention. In addition, as I will later tell, I was called to meet with Lyndon Johnson to talk of the world he now faced, and, along with Ted Sorensen, who wrote the major share, I worked on the speech L.B.J. was to give to the Congress. I also drafted a special obituary notice for the *Washington Post.* (I recently reread it, a miserable piece of turgid drivel.) A very compelling task was to decide who could come to the funeral; after the family, the Cabinet, the Congress and the visiting heads of state, there was little remaining space.

Such was the intensity of effort that it took thought away from the compelling sadness of the day. Only later did I come fully to reflect on what we had lost. It was different from Roosevelt's death. His loss meant a world come to an end. The loss of Kennedy was that of a well-loved friend. Life went on.

Since that sad November day, the comment on the Kennedy years has moved in two radically opposing directions. First have been chronicled the achievements of the young President, the grace, prestige, intelligence and public interest he brought to the presidency, his calm resistance to military and Cold War excess. The result was and remains a generally admiring view of John F. Kennedy and his domestic and foreign policies. But there has been another and highly aggressive discussion that has centered on J.F.K.'s health and his extramarital sex life. This, but especially comment on his sexual activities, has supported a sizable and well-rewarded literary community.

Of Kennedy's health problems I was broadly aware. In outward aspect, he was vigorous and alert. We all knew, however, that he was often in pain, back pain in particular, and he had recourse to a range of drugs that eased the suffering, making tolerable his public life. I had heard of his Addison's disease without ever being quite clear as to its effect. All of this, with varying degrees of emphasis, has now become widely known.

More celebrated has been the Kennedy sex life. Of his sexual adventures, real or alleged, I never heard in all the years of my association with him. They were never discussed in my presence. Rarely is it possible to plead ignorance with more satisfaction. All of my knowledge, such as it is, was acquired after his death.

That John F. Kennedy had no extracurricular affairs, I would not suggest. Franklin D. Roosevelt did; so, by his own sometimes enthusiastic assertion, did Lyndon Johnson. There were Grover Cleveland and others, going back to Thomas Jefferson. And now Bill Clinton. However, upon Kennedy, for a full third of a century after his death, major public attention has been fixed. The reason is reasonably clear: it is not that J.F.K. was exceptional; rather, it is that on him the interest is still centered, and it is from this that the money comes. It is also true that sex is something that even the dimmest commentator can understand, the least literate can write about.

Kennedy as President, as, indeed, all modern Presidents, did not govern alone. There was his intimate circle of protectors and assistants, notably the Irish Mafia already mentioned. There was his brother Robert Kennedy, intelligent, determined, energetic, on whom there will be more. Also, though in a diminishing and soon diminished role, his father, Joe Kennedy, Sr. There was the larger loyal and talented Kennedy family, J.F.K.'s sisters and his brothers-in-law, especially Sargent Shriver. And, most of all, there was Jacqueline Kennedy. To this inner Kennedy circle and to her I come in the next chapter. But first a word from her on her husband and the effect of his early death on his enduring reputation.

Jacqueline Kennedy and I were standing some months, perhaps a year or more, after the President's assassination on a lawn across the Charles River from Harvard—down river a bit from the Harvard Business School, more formally the Harvard School of Business Administration. It was then a possible site for the Kennedy Library, not an attractive one, for the dominant view would be of the belching towers of the Cambridge Electric Light Company.

"My children, as they grow up," Jackie said, "will have difficulty believing that Jack, so young, could have been President." Then she added, "They would have had a different view had he become an old man."

She made a powerful point. Had Kennedy lived, his reputation would, indeed, have been very different. There would have been the immutable effects of aging and age. He would also have been a forced participant in the discussion of the disputatious events of his presidency—the Bay of Pigs, the Missile Crisis and, inevitably, Vietnam. Better to have these left to the historians.

Worse still, he could not have avoided the eager discussion of his sex life. All this he escaped.

The murder of John F. Kennedy brought sorrow to those who loved him. And to the country and the larger world. It had one little-mentioned effect: it left us with the enduring memory of a young man at the very height of both his political power and his personal charm.

9

The Kennedy Circle, Jacqueline Kennedy

IT IS SAID that when Woodrow Wilson was President, he tapped out his speeches and other communications on his own typewriter; he had no secretarial staff. As the twentieth century passed, however, the President became not a person but an institution. He governed; he also presided. So it was with John F. Kennedy.

In the White House during his administration, or rather in its prestigious West Wing, was the indispensable Theodore Sorensen. Also Lawrence O'Brien, Kenneth O'Donnell, Myer Feldman and Richard Goodwin, all of whom had been with J.F.K. in the Senate and on numerous campaigns. Together, they handled congressional relations, the mass of information and requests on domestic matters coming in from government departments and agencies. Most of their work was agreeably unknown. Present there as well were the far more controversial foreign-policy advisers, on whom a later word.

And especially remembered by all was Kennedy's long-time friend, the Oval Office gatekeeper, David Powers. A greatly underestimated figure, he combined rich intelligence with an unfailing sense of humor. Alive in my memory is a much later encounter with him—the day, years after the President's death, when the Kennedy Library was opened on Columbia Point in seaside Boston. It was an occasion of nostalgia and grace. A soft wind blew in from the ocean; President Carter was there, as also Jacqueline Kennedy Onassis, the Kennedy family and a large assemblage from the Boston political establishment. The speeches were short, engaging, affectionate. I walked away afterward with Dave Powers and made a routine comment on what a good day it had been. Back came his reply: "Ken, never in the history of Massachusetts Democracy have so many gathered so happily with so few under indictment."

Also in the White House, as Special Assistant to the President with an intensely important advisory role, was Arthur M. Schlesinger, Jr. He brilliantly helped shape the history of which he would later write. His was a voice of sanity on the two great misadventures of the Kennedy years, the Bay of Pigs and the enduring tragedy of Vietnam. For them, the grave misguidance of the foreign-policy advisers was responsible, the error extending out to those in the State Department, in the CIA and especially in the Pentagon.

Finally there was Robert F. Kennedy, who headed the Justice Department as Attorney General but who was, in fact, a main source of White House policy guidance. John F. Kennedy did not readily rebuke error or dissent; Bobby was under no such constraint.

In recent years, in the presidencies of Ronald Reagan, George Bush and, as this is written, of Bill Clinton, the Washington press, radio and television corps has been nourished by a constant flow of information on the wars within the presidential staff and the conflicting aspirations—who is up and who is down; who is in or on the way out; what executive and congressional action is favored by the contending forces; who might resign or be sacked; what book a departed statesman will write and, more particularly, at what price. There was all but none of this during the Kennedy years; at most, and rarely, there was a word on differences in personal view. What is now commonplace would then have been thought exceptional, unpleasant and overtly disloyal.

Much of this peace and tranquillity can be attributed to Robert Kennedy. He made it painfully clear that public criticism or dissent from established presidential policy was not permitted; it was beyond the pale. And his discipline extended well outside the immediate circle. In the aftermath of the Bay of Pigs debacle, Chester Bowles, then Under Secretary of State and a critic of that visibly insane enterprise, was warned to silence, his coat lapels tightly held by Robert Kennedy. Others got the message and, as did Schlesinger and I, made no public mention of their opposition.

*

During the presidency of Jimmy Carter there was a certain adverse effect from the President's brother, Billy; his amiable entry into beer-drinking contests and similar athletic events had a less-than-favorable resonance. A President needs the full support of a well-behaved family. This was and, had he lived, would have continued to be the good fortune of John F. Kennedy.

But not completely. There was Joe Kennedy, Sr., known universally in those years as Old Joe. His far-from-retiring personality, his unduly talented money-making, his Hollywood entanglements and his singular ambassadorial history all cast a shadow. In the 1960 campaign he was still a formidable presence, although not always a conservative voice. In one intense discussion of a prospective economic-policy speech to business journalists in New York, what was to be a major economic statement, Joe Kennedy supported my generally liberal views. Others, including his son, were more restrained. Afterward, only half in amusement, J.F.K. said to me, "I'd like to know what gave you this hold on my old man." But Joe Kennedy was fading from the scene. He faded more rapidly after his son came to the White House and completely when, in 1961, he suffered a disabling stroke. His sons visited him at Hyannis Port, took visitors to see and talk to him, but his influence was over.

Robert Kennedy was known to have his own agenda; far more than his brother, he was identified with the social concerns of the poor and the deprived. He had also an independent interest in a wide range of other matters. In pursuit of information and policy guidance, he convened regular meetings, seminars, at his house in Virginia. Problems and needed action, foreign and domestic, were discussed and debated by acknowledged or self-avowed experts. The Hickory Hill Seminars were a hallmark of the Kennedy years. No one expected anything quite like them under Ronald Reagan.

A basic point about Robert Kennedy remains: he was the principal moral force both in the White House inner circle and the larger administration. In the Washington political community in those years John F. Kennedy was loved; Robert Kennedy was more than slightly feared.

The President's sisters were also very much visible, with strong views on social issues, and two of them, Eunice and Jean, brought highly effective husbands, Sargent Shriver and Stephen Smith, into what was everywhere called The Family. Sargent Shriver became the founding head of the Peace Corps and, as I will later tell, remained with Lyndon Johnson to help design and lead the War on Poverty. I have a special memory of his talent in diplomatic persuasion.

He came to India in my early days there as Ambassador to persuade the Indian government to accept the very first of the Peace Corps contingents. I was not at ease on the matter. I warned him that in the Indian mood of the time, and that of Jawaharlal Nehru in particular, the Peace Corps would be regarded as a rather obvious example of the American search for influence—what was seen in those post-imperial days as the American threat. It would have to be presented as small, experimental and confined to one area of India, the Punjab. In the meeting with Nehru, Sarge, as he was known, followed my instructions carefully, but he did so with earnest eloquence. The new project he headed would help the most needy of the Indian needy; it would give American youth experience in good works. Nehru listened without comment. I was prepared for the worst. When he eventually replied, it was to ask why the enterprise had to be so small, why it had to be limited to only one Indian state. He thought the idea excellent, regretted only the evident limitations. My reputation for wise guidance was visibly diminished.

The other members of the family had a similar social commitment. Eunice Shriver began a lifelong effort in those years on behalf of disabled and handicapped children. Jean Kennedy Smith was a quiet presence; she would one day be a far-from-passive Ambassador to Ireland. Her husband, Stephen Smith, handled the considerable financial resources of the family and did it so well that its members, including the President, remained untouched by any breath of scandal or even any political comment. Given the dollar amounts involved, this was no small achievement. Some political roles are important because they are unobserved.

In the late autumn of 1960, after I had made clear my interest in India, it briefly crossed my mind that I might succeed Kennedy as senator from Massachusetts. I spoke to the President-elect about it. It was a question of where I could do the most good—or so, inevitably, I put it. Kennedy was very clear; he said it was in India. I later learned that the family succession had already been planned. The youngest brother, Edward, was too young to be appointed to his brother's seat at the time. Two years later, however, when Ted reached thirty, the temporary appointee would retire according to a well-understood plan, and Ted would be elected to the Senate in his own right.

Edward Kennedy's age and youthful inexperience were, at the beginning, sadly adverse circumstances. But not always. Arising early one morning in his first campaign to shake hands with the workers arriving at a Massachusetts factory, he was greeted by a man of mature years who came rolling down the line.

He said, "Teddy, m'boy, I hear you've never done a day's work in your life."

It was the candidate's most vulnerable point; he braced himself to make a reply, but the old man didn't wait for it: "Let me tell you somethin', lad. You haven't missed a thing."

J.F.K.'s decision on his successor was better than he could ever have imagined. Edward Kennedy was elected to the Senate in 1962 and went on to become the most diversely effective liberal legislator of his time.

I now turn to the best-loved member of the Kennedy circle and, for many, perhaps the least understood.

*

Jacqueline Kennedy was beautiful, with a beauty that attracts and holds the eye. She was impeccably—it was said expensively—but never ostentatiously dressed. Her greeting on encountering you, low-voiced, warm, had the effect of a physical embrace. You were, without any evident exaggeration, the person she truly welcomed and most wanted to see. As no one in the Kennedy circle spoke ill of the President, certainly no one spoke ill of his wife. There were occasional differences between Jackie and her numerous sisters-in-law as to manner, style and political expression, but they were not carried to the larger public.

Nor was the reaction from the larger world ever less than positive. First Ladies, that offensive term, have, since the Kennedy years, come in for a fair amount of criticism. This, they invited as they shared or assumed a public role or, as in the case of Nancy Reagan, took over from an aging and, toward the end, mentally detached husband. Any serious attack on Jacqueline Kennedy would have been dismissed as idiosyncratic.

This approval was attributed to the way she stood apart from presidential politics. She decorated, furnished and otherwise refurbished the White House. She was elegantly present on all social occasions. It was assumed that this was all she did; she was not politically involved. For many, then as now, this was how it should be for a President's wife.

There could have been no greater error: Jacqueline Kennedy *was* deeply involved. It was only that with acute self-knowledge she had chosen her own role carefully. She did what no other person in the Kennedy circle could do as well, and what she did occasioned no serious public discussion.

Jacqueline Kennedy did not talk politics; she celebrated her detachment therefrom. One of my early recollections is of breakfast at the Kennedys' house in Georgetown during the campaign of 1960. J.F.K. and Jackie were leaving for a day of vote-getting in Pennsylvania. He had a briefcase full of speech drafts, political memoranda, biographical material on the politicians he would meet. She had the *Mémoires du Duc de Saint-Simon* in French. She would take no part in the day's political persuasion. It was only important that she be there; she, as much as the candidate, was the one the crowds wanted to see. But she had a deeper purpose: it was she, not the more trusting J.F.K., who would observe, hear and render judgment on the politicians they would encounter.

The basic and commonplace problem of any President is how to react to a new situation, foreign or domestic, great or small. But this evokes further and slightly less evident questions. On whom can he depend for guidance; whom can he trust? The President cannot make all the necessary decisions from his own knowledge alone. Here was the role of Jacqueline Kennedy in her husband's administration. She knew and observed the people on whom the President had to rely; she distinguished sharply between those who were serving him and those who were serving themselves. And especially those who concealed imperfect judgment behind a display of personal importance—the accomplished frauds.

Her wise and astringent analysis was especially important to John F. Kennedy. He was, by nature, a tolerant person. His years in politics in both Massachusetts and the Congress had made him open to a great number of individuals who, along with their advice and service, had their own personal purpose to advance. When he became President, the need to know on whom he could depend became even more urgent. Jackie's view of people was essential. She made no conscious decision to analyze them; she simply took for granted that it was her job. When she saw a basic commitment to self-advancement, some general evidence of inadequacy, she made it known as a quiet statement of fact.

In those days General Lyman L. Lemnitzer was the Chairman of the Joint Chiefs of Staff, for the President a critically important post. Lem, as he was known, was an agreeable man of impressive military aspect. He was not given to deep thought. "Jack thought well of him," Jackie once observed, "until one Saturday morning he came into the White House in a sport jacket." She had already seen that it was the uniform that held Lem up.

John F. Kennedy's short life came to an end on that autumn day in 1963; for another thirty-one years Jackie remained one of the best-regarded women of her time. Partly, without doubt, this was the Kennedy legacy. She inherited the imagery, the affection, that had graced the Kennedy years. He was gone; she lived. She remarried but remained very much her own person. She continued her association with the Kennedy family, which was by no means casual. In 1968, that most contentious of years, as I will tell, I helped initiate the effort to persuade Eugene McCarthy to run for the presidential nomination against Lyndon Johnson and specifically against the major tragedy of the Vietnam war. When Robert Kennedy came into the race that spring, I remained with Gene. Jackie was outraged. She rebuked me sharply for my defection from The Family.

There was more to Jacqueline Kennedy's role than her judgment of people and her strong sense of loyalty. She also listened to and charmed politicians, domestic and foreign, from Charles de Gaulle to Jawaharlal Nehru. During her wonderfully agreeable visit to India in 1962, Nehru insisted on moving her and her sister, Lee Radziwill, from the rather commonplace house we had borrowed for her visit—the embassy dwelling was then under construction—to a more spacious and decidedly more attractive apartment that opened on the great lawn of his official residence. He told her with pleasure that these were the rooms once occupied by Edwina Mountbatten (the wife of Louis Mountbatten, the last Viceroy of India), with whom Nehru had an association the nature of which has been for long, if cautiously, debated.

Jackie's attraction for other political leaders, including

even Khrushchev, is part of the history of the time. The list included congressmen, senators and high public officials who came to the White House. No one ever paid more evident attention to what they wanted to tell of their own achievements, and nothing so attracts a public figure as a genuinely receptive audience. Of this, Jackie was well aware.

My friendship with Jackie became even closer in the terrible days following Kennedy's murder. There were the funeral arrangements, over which she took command and on which I helped. There was the far-from-trivial matter of her immediate and future life to be discussed—along with other vexing problems, the question of where she and her children would live. I arranged that Averell Harriman evacuate his spacious, well-staffed Georgetown house and Jacqueline Kennedy, Caroline and John, Jr., be installed therein. The Harrimans went to a hotel. Later, Jackie and I met at Hyannis Port, for a family skiing expedition in Switzerland, during a memorable visit to our place in Vermont and frequently in New York. Then came her last, immeasurably sad days, the wake and her funeral. It was during the long years of our friendship that I came fully to understand what she had done for John F. Kennedy as politician and President. And what made her the most admired woman of her day.

10

Jawaharlal Nehru

THE FOREIGN POLITICAL leader with whom I had the closest personal relationship over the years, apart from Roy Jenkins, who is yet to be mentioned, was Jawaharlal Nehru. He was the Prime Minister of the newly independent India when I first knew him. He had been much more. With Mohandas Gandhi, he had led in the long battle to achieve his country's independence, including, alas, its division into three, often hostile segments. And to establish the government of the new India and shape its economy and polity. With Gandhi, Jawaharlal Nehru was, indeed, India: Gandhi was its history; Nehru, after independence, its reality.

I first met Nehru in the 1950s, when I was recruited as a less than major adviser on India's Second Five Year Plan. Then, as American Ambassador for two and a half years in the early 1960s, I was in constant association with him. At our first meeting after I became an alleged diplomat, Nehru told me, to my delight, that he hoped my new status would not keep me from continuing in my earlier advisory capacity. Few ambassadors have been so favored, even in a ceremonial gesture.

There was a further element in our relationship: we both had been at the University of Cambridge, to which I had gone in 1937 to study under John Maynard Keynes (who, unfortunately, was ill that year) and where Nehru had been as an undergraduate. We were both associated with Trinity College, at which he studied and of which I later became a Fellow and an Honorary Fellow. We also shared a common commitment to the thought and advocacy of Sidney and Beatrice Webb, Harold Laski and, more generally, of the British social left. Nehru made no secret of his British background and its influence on his political thought. He once said, "You realize, Galbraith, that I am the last Englishman to rule in India." In Washington, on a state visit in 1961, he expressed amused discontent that Senator William Fulbright, Secretary of State Dean Rusk and one or two others had been Rhodes Scholars—"so many Oxford men." Then, relaxing with a smile, he added, "However, the world still belongs to Trinity men."

Nehru's British university connection was not something to be taken lightly. In the early years of the century, Britain was the world center of democratic socialist thought. To Nehru, as to many others, this was received truth. Here was the path to the political and economic future.

This being the true way for Nehru, it became so for independent India. The economy would reflect what was called the socialist pattern of society, the word "pattern" allowing for a certain flexibility of choice. Soviet Communism was too comprehensive, and there were also—a less mentioned point, alas—the liberties of the citizen to be considered. But capitalism was unacceptable as well. So India would stand apart from Soviet-style socialism, but she would also be aware that she might be a target of capitalist imperialism. That, after all, had been her past experience with Britain. So something in between was best. In the 1950s these ideas sustained an intense discussion in India in which I was much involved.

With others, I was attracted by the idea of a new social system in a new country; one would have been dull to react otherwise. However, I had doubts as to socialism. It meant, of course, public ownership and operation of industrial enterprises. These might well become extensions of the public bureaucracy and subject to political pressures, politically motivated appointments and the general lassitude of public agencies. Sensing that, I wrote a paper warning against what I called "Post Office Socialism"; it had a wide circulation, was debated in the Indian Parliament, and the reference passed into the language. Later I also became aware of the adverse effects of government regulation. Supervision and control were the obvious answer to an otherwise uninhibited and dominant capitalism. And, especially as regards foreign investment, the evident manifestation of capitalist imperialism. In practice, however, such regulation meant delay and frustration in economic development and the creation of a small new industry in which public servants, so described, sold the permissions necessary to build or invest. In the early 1960s, when I was Ambassador, I had my longtime and remorselessly conservative friend Professor Milton Friedman to lunch one day at the Embassy Residence. My views had sufficiently changed that I was moved to tell him (and the other guests) that India was one of the two countries in the world (the Soviet Union being the other) that he could visit to their advantage.

The concept of a special type of economy and society, neither fully free enterprise nor completely socialist, remained strong in Nehru's mind until his death, in 1964. It did not survive him. Liberalization, the freedom from restrictive controls—from the need for the government permissions—became accepted in public rhetoric and, if less so, in practice.

These were also the years of the Cold War and the arms race, and Nehru had a larger role to play. In the spirit of Gandhi, he managed to stand above both—and was sought out, in consequence, not only by the United States but by the Soviet Union. No other leader of a less-developed country attracted such attention. This privileged position was accorded his successors, only to be ended by a mindless, useless and even disastrous recourse to the testing of nuclear weapons. India became a minor player in the nuclear arms race.

While I was Ambassador, my association with Nehru was at two levels, official and personal. Officially, there was the need to assure him that the United States could be a good friend of India's, extend generous assistance without having imperial ambitions. Persuading him that this was Kennedy's position and my own was not too difficult; the problem came from the traditionalists in the State Department, who were still locked into the Dulles military diplomacy and were now joined by the enlightened cold warriors, as they saw themselves, of the Kennedy Administration. Especially troublesome was the military alliance they promoted with Pakistan, along with the weapons support that went with it—Pakistan as a bulwark against the Red Army! I managed to soften somewhat the troublesome message this sent, but it was a less-than-agreeable aspect of my public association with Nehru.

Other matters were more pleasant, including Nehru's occasional offers of advice to John F. Kennedy—the older experienced statesman to the young President. These were well-intentioned and relevant. Especially important was his guidance on Nikita Khrushchev, with whom he had had practical experience. There was no necessary connection, he wanted Kennedy to know, between Khrushchev's speech and his action. Once, when the Soviet premier mounted a major verbal defense of the Berlin Wall while in India, Nehru told him he didn't himself think much of the East German government it protected. Khrushchev replied that neither did he. At Nehru's suggestion, I passed this along to Kennedy just before the next American-Soviet meeting.

On occasion Nehru had a critical word on my published views. He had read my *The Affluent Society;* he thought I was too easy on the society I there described. Technological innovation was not, as was accepted in the book, a measure of human advance. Only two modern inventions he thought clearly contributed to social well-being: one was the bicycle; the other, electric light. In modern Delhi some tons of pollutants are now dumped into the atmosphere every day, much of them from motor vehicles. The traffic jams outdo those of Manhattan. Nehru's foresight may have been better than I then imagined.

This, in small part, was our official relationship. It was from our personal association that I came to appreciate and, to some extent, understand the Nehru personality. He came often to our residence and we to his. He was attracted to Catherine Galbraith, as to other intelligent and engaging women. It is this informal contact on which I now look back with the greatest satisfaction.

Jawaharlal Nehru was slender, of slightly less than average height. Yet he never seemed a small man. He wore a tweed or worsted coat that buttoned up to the neck, a fashion that came to be known around the world as a Nehru jacket. Of a family originally from Kashmir, he was light-complexioned and could pass readily for a European. The farther south one goes on the subcontinent, the darker the prevailing skin color; the ancient effect, most believe, of the ever-more-relentless sun. Nehru's family had been a formidable force in India, his father a prominent attorney and an early advocate of Indian self-government. Nehru's daughter, Indira, and her son Rajiv were to follow him as Prime Minister. His sister, Vijaya Lakshmi Pandit, known to all as Nan, was, at various times, Ambassador to the Soviet Union and to the United States, President of the U.N. General Assembly and a voice on a wide range of foreign-policy matters. She was also a close personal friend of our family. As was and is Nehru's cousin B. K. Nehru, who was also at one time Ambassador to the United States and Governor of Kashmir.

Jawaharlal Nehru had a warming sense of humor and appreciated it in others. One example remains in my mind. As might be expected, little from the United States was more diligently reported in the Indian press in those days than the matter of our aid appropriations, including their progress through the Congress. Once, Congressman Singh Saund, of southern California and Sikh origin, held up the foreign aid bill while pursuing some personal political purpose. A large share of the money being discussed was intended for India, and the Indian newspapers headlined the delay. On the day the story appeared, Nehru came late for an appointment; there had been lengthy negotiations on the enduring problems of the Punjab. He apologized and, somewhat uncharacteristically, added, "These Sikhs [the dominant force in the Punjab] are very difficult people." I pointed to the front page of the newspaper on the table. He reacted with genuine delight. "One Sikh," he said, "can hold up the entire government of the United States. I have forty million of them."

On another occasion, my wife and I had as our guest Angie Dickinson, the film actress and a longtime friend. She was then at the summit of her career, combining beauty, intelligence, political interest and general charm. As did all visitors to India, many of whom I had to restrain, she wished to see Nehru. One afternoon I sent him a note, saying I knew him to be busy but could he spare a moment for a lovely Hollywood star who would like to meet him? Within the hour I had word back saying that in great emergencies he could always make time. Could I bring my guest over at once? I found Angie, took her to the Prime Minister's residence, and they talked for nearly two hours. I remember especially one question from Nehru. "Miss Dickinson, when you are making a film, you spend some months studying and then creating the character you are playing. Doesn't that have some permanent effect on your own personality?"

To Nehru's delight, Angie replied, "I certainly hope not, Mr. Prime Minister. In my last four films I've been a woman of deep ill-repute."

I might note in passing that Nehru's sense of humor extended to enjoying that of others, including Gandhi's. An example he once quoted to me occurred when Lord Irwin was Viceroy. (He was later, as Lord Halifax, exported by Churchill to the United States as Ambassador.) Predictably, he and Gandhi had serious differences of view. One day at Gandhi's ashram in Ahmedabad, a friend and supporter sought to ease the conflict by saying, "Mahatma, you must know that Lord Irwin never makes a decision without praying over it first."

Gandhi reflected on this for some minutes. Then he said, "And why do you suppose God so consistently gives him the wrong advice?"

Politicians in high position must usually justify themselves, show that they deserve the eminence they enjoy. Nehru took his role fully for granted, not only in India but in the world at large. I heard him give public speeches; several times I heard him speak in Parliament. Always his leadership was assumed, not defended. He never asked for support; he believed it was already his. He never had a script; he simply told what was on his mind. That was what all should know.

He had, more than incidentally, the advantage that Indians have an enduring appetite for oratory. In Western democracies, those listening to a lengthy speech think of the more interesting things they might be doing; they are hoping for the end. In India, a public address is a welcome alternative to idleness, even greater boredom. (The often appalling Indian motion pictures serve the same function.) Accordingly, the longer the speech, the more it is appreciated. This, Nehru fully understood.

In private discussion with other officials, including ambassadors, Nehru did not respond if he did not agree or he had no opinion. He simply remained silent. In the autumn of 1961, during his first formal meeting with John F. Kennedy, the President urged various matters, including peace with Pakistan, a solution on Kashmir, something on economic development, the other issues of the day. Nehru made almost no reply throughout. Kennedy was deeply disturbed and later called me to tell of his concern. I explained the situation; a second session did go slightly better.

I was not unprepared. Earlier that year Lyndon Johnson, as Vice President, had come to New Delhi. During a long morning with Nehru, there was no substantive response to any point Johnson raised until, toward noon, when Nehru joined L.B.J. in approval of rural electrification. It fell to me to compose the press communiqué telling of the accomplishments of their meeting. It was perhaps then that I discovered my talent for fiction, which led on to several well-regarded novels.

When I was leaving India in 1963 and said farewell to Nehru, it was evident that his life, both physical and political, was over. More painful than his health problems was the political disaster that had begun for him in the autumn months of 1962. After some minor border maneuvers by Indian troops, the Chinese launched a major offensive against India on the high Ladakh plain adjacent to Tibet, the least edifying non-Arctic landscape anywhere on the planet, and to the east down the Himalayas and into populous Assam. The war received little notice in the United States, even in Washington, for it coincided almost to the day with the Missile Crisis in Cuba.

The handling of this Far Eastern adjunct of the Cold War, as Secretary of State Dean Rusk and other cold warriors saw it, was formally delegated to me. It brought me into daily, sometimes hourly, contact with Nehru. Military transport and arms aid had to be air-lifted from Europe. More important, a drastic reorganization of the Indian military leadership was needed, for both the civilian and military officials were deeply incompetent. And there was much more to be done.

Through all of this Nehru was visibly depressed. His vision of India as standing above such cruel and useless conflict was at an end. The country, and more particularly the politicians, were showing signs of being militarily aroused. As this occurred, the Chinese early one morning offered a ceasefire. There was no time for instructions from Washington; I went to see Nehru to urge that he accept it. Tired and worn, he immediately agreed; it was the worst time of his life.

In the view of the State Department scholars, however, it was the best time, when Nehru was weak and defeated, to press him on Pakistan's claims against India and especially the much-disputed status of Kashmir. At the very moment when Nehru needed to show political recovery and strength, we would advertise and exploit his weakness. The British made matters worse by joining the effort and adding to their delegation in New Delhi the two most unfortunate figures in modern British foreign policy: Winston Churchill's formidably arrogant son-in-law Duncan Sandys and Lord Louis Mountbatten. (Nothing in the twentieth century was so badly handled and with such disastrous consequences as Mountbatten's policies on Indian independence, leading as they did to the division of the subcontinent into three countries amidst conflict, mass migration and death.) Our role was eased by the participation of Averell Harriman (on whom there will be word in a later chapter). He became the subject of a memorable Nehru comment: "I will negotiate with Harriman, because he is a gentleman. I will not negotiate with Duncan Sandys, because he is a cad."

The talks with Pakistan ground on for many weeks; nothing was accomplished. I was more acquiescent in the process than I should have been. A comment I made at the time, however, still has resonance: "In the present state of incompatibility, these efforts lead not to pacification but only to more combat."[1]

Sometime in the summer of 1963, I met with Jawaharlal Nehru for the last time. He was old and tired; the confident manner that allowed him to believe he was the designated leader of all the Indian people was gone. In little more than a year came the news of his death. India and the world had lost a major force for civilized values. I had lost a truly admired friend.

The State Department dispatched a delegation of mourners to his funeral that included those who, urging arms to the subcontinent, had contributed most to his despair.

1. *Ambassador's Journal* (Boston: Houghton Mifflin, 1969), p. 575.

11

L.B.J.

IN HIS meant-to-be-secret and widely published White House conversations on the several actions he was urging against decency and the public interest, Richard Nixon regularly referred to himself as R.N. With this appellation, he hoped to place himself with F.D.R., J.F.K. and L.B.J. It didn't take, perhaps because there was no readily resonant middle initial. M for Milhous was somehow awkward, not compelling. More important, being known by initials is an indication of affection; this Nixon, in singular measure, was not accorded.

The initials L.B.J., in contrast, stand as one of the best-recognized designations of modern times. So Lyndon Baines Johnson is denoted. The library and educational center that were created in his memory at the University of Texas are rarely given their full names; they are the L.B.J. Library, the L.B.J. School.

In politics the most common, indeed the most commonplace, distinction is between the man or woman who holds public office in order to enjoy the personal pleasure it provides and the one who sees such a position as an opportunity to effect public action and change. The first is forgotten; the second becomes the history of the time. In recent years George Bush, in a tradition going back to Calvin Coolidge and before, was one for whom enjoyment of the office was a principal motivation. This was well recognized; Bush, in consequence, is lost to fame. Lyndon Johnson, L.B.J., was at the other extreme; for him, political office, while indeed greatly enjoyed, was for what he could accomplish. Kennedy always used less power in pursuit of his goals than his position and personality provided; Lyndon Johnson always used more.

There was even a physical aspect to this effort. Anyone whose action was needed by L.B.J. might be firmly grasped, shaken slightly, while being told what he must do and when he must do it. The only escape was acquiescence and compliance. This was a small symbolic manifestation of how Johnson dealt with the Congress and, indeed, the electorate as a whole. Back of it was Lyndon Johnson's commitment to the public good and his belief—more precisely, his faith—that government was made to serve it. This conviction, however, was centered almost exclusively on domestic-policy matters. He had less certainty as to the role of the United States beyond its borders. This, in turn, made him vulnerable to the most disastrous advice on foreign and military policy that any American President was ever accorded. It was the ultimate tragedy of Lyndon Baines Johnson.

*

Johnson's self-assurance on domestic policy was the product of both his personal history and his political career. The first began in a family of modest income and aspiration in Texas. He did not suffer deprivation as a child; his youthful experience of poverty was, however, never understated. He graduated from a small, little-known Southwestern college, first emerging in public life as a local official of the National Youth Administration in Texas. This agency had a special responsibility for giving work and income to the disadvantaged, and it was there that Johnson began a lifelong identification with the underclass and its problems. The poor became his particular constituency as he envisaged it. So it was when he went to the Congress, and so it remained as he went on to become Vice President and President. On these matters his personal judgment was controlling and good. On the larger world scene he accepted the guidance of his advisers. He assumed they were right—a grave mistake.

As earlier noted, I came to know Lyndon Johnson in the early 1940s. We were Roosevelt liberals, each in his thirties, serving in Washington, he as a member of Congress, I as the deputy administrator of the Office of Price Administration in charge of wartime price control. One of our first meetings, perhaps the first, was in the home in nearby Virginia of the Clifford Durrs. Durr, a greatly respected lawyer, had served in several different New Deal agencies. He and his highly motivated wife, Virginia, were both from Alabama; both had a passionate commitment to civil rights, a deeply felt but exceptional response to their life in the South. In later years the media and the electorate expressed a measure of surprise that L.B.J., a Texan, should lead in the battle for voting rights and against the various forms of racial segregation and discrimination. Knowing of his friendship with the Durrs, and recalling our earlier discussions, I understood.

As the years went on, I saw Johnson socially in Washington, on Capitol Hill, at his ranch in Texas, when he was Kennedy's Vice President and then in the Oval Office. In the Roosevelt and again in the Kennedy Administration there was a strong infusion of Eastern academic talent into the Washington governing community; like Roosevelt and Kennedy themselves, many of those thus recruited had been at Harvard. With them, Johnson was always slightly uneasy. I was one of them, but in earlier times, as I've said, I had been much committed to agriculture; I began my academic life as, in the common reference, a farm economist, and I had been born and raised on a farm. This gave me a standing with Johnson that was not enjoyed by my university colleagues; we were both farm boys.

On occasion our association did require some dissembling on my part. Once while he was Senate Majority Leader and was still recovering from his first heart attack, Lyndon, as I then knew him, took me dove hunting in the Texas hill country. It was a lovely day. We rode in separate jeeps to the hunt. I was handicapped by the fact that rarely before had I shot a gun at a stationary target and never at a bird in flight. When a dove came in view, I pointed the gun in a general way, shut my eyes and pulled the trigger. Needless to say, the bird escaped, as fortunately doves often did. Lyndon, as we rode back to the ranch, allowed me to believe that he had succumbed to my fraud.

Any measure of contrivance I might attempt was, however, slight as compared with Johnson's. In this, he was supremely adept. I particularly remember a time when I was back from India on a short visit in the early sixties and went to Capitol Hill for some committee testimony. Afterward, I met Johnson near the Vice President's office, and he asked me in for a chat. (A Vice President must learn to suffer frequent periods of idleness; this was a common affliction for L.B.J. under Kennedy.) Engaging in reminiscence, I asked him what it had been like to be a senator from Texas and have had to deal with the Texas oil lobbyists. The independent oil producers had been particularly ruthless in their pressure on me as a price fixer. I asked if he knew one unusually rapacious lobbyist, whose name I've long since forgotten.

"Did I know that bastard?" L.B.J. responded with Johnsonian emphasis. "He came into my office one day and said, 'Senator, you have a tough race coming up. If you can do better on a few things like the depletion allowance, we can come up with ten thousand dollars for your campaign.'

"I tell you, I really told him off. 'You can't come into the office of a U-nited States senator and talk that way.'

"He never moved. 'We could make it fifty.'

"That really got me. I told him he was talking downright bribery.

"He didn't budge but said, 'Senator, you have the position, the experience, the influence. Let's talk about a hundred thousand.'

"That was it, Ken. I called in Walter Jenkins [a longtime Johnson assistant]. He took one arm, I took the other, and we marched that fella right outa my office. And I said, 'You stay out of here, you bastard. You're getting too close to my price.' "

That incident almost certainly never happened. With much else, Johnson was accomplished in using fiction as a cover for plausible fact, and his talent extended to an inspired use of metaphor. In response to a suggestion that someone once made that he appeal for help on a needed piece of legislation to a former colleague in the Senate, his rejection was recited all over Washington: "That fella has the same effect on political action as pantyhose on finger fucking."

A particularly memorable example of this gift for metaphor came on a summer day in the mid-sixties, when we had already split over Vietnam. A call came to our place in Vermont. "Everyone around here is worn out," Johnson said. "I have a couple of speeches pressing me. Let's forget these foreign problems; come down and give me a hand." I agreed, and he sent a plane to Keene,

New Hampshire, to bring me to Washington. At the White House I labored through a long afternoon in the office of his assistant, Jack Valenti, adjacent to the Oval Office. (Valenti in his memoirs has also told of the occasion.) One speech was unimportant, a protocol exercise for some diplomatic occasion. The other was to be a major statement on economic and social policy. In this, I put Johnson's beliefs, as I knew them, as many of my own as he might be persuaded to accept and such rhetoric as was needed to carry the message.

By late in the afternoon, I was finished. Johnson came into the outer office, put his foot on a chair, looked at and set aside the unimportant speech and settled down to the major address. He nodded and smiled; to my pleasure, it was clear that he liked it. When he was through, he said, "Ken, you've saved my life. It's exactly what I want to say. I'm not going to change a word." This is the mark of a secure person. Most would say, "It's pretty good; I'll only make a few revisions."

Johnson's face then saddened. "It's good, but nobody else will think so. Did it ever occur to you, Ken, that making a speech on ee-conomics is a lot like pissin' down your leg? It seems hot to you, but it never does to anyone else."

Never since have I given a speech on economics without having that metaphor in mind.

Another L.B.J. episode also remains with me. One day not long after the great 1964 victory over Goldwater, I arrived home from an exceptionally long day at the university. We had an evening engagement. I asked Emily Gloria Wilson, short, intelligent, black and for forty years our loyal housekeeper with a strong commitment to our children, to hold off any telephone calls; I needed a rest. Shortly thereafter L.B.J. called. As was often his custom, he was on the line himself.

"Lyndon Johnson here. Get me Ken Galbraith. I want to talk to him."

"He's resting, Mr. President."

"Well, get him up. I need to talk to him."

"No, I'm sorry, I can't. I work for him, not for you, Mr. President."

Later, when I was awake and heard the details, I was not amused. I promptly called back to make amends. L.B.J. came on the line; it was pure Johnson: "Who is that woman who works for you? I want her down here in the White House."

There was another occasion when, despite our differences on the Vietnam war, we were briefly in touch. It quite possibly changed the course of Greek history. I have often told the story, as did the central figure himself, Andreas Papandreou, who for many later years was either the Prime Minister of Greece or the dominant contender for the post.

On the evening of April 21, 1967, came word or at least an impression that Papandreou, who was in prison, was to be shot by the ruling Greek junta, the colonels, to whom he was bitterly and effectively opposed. Papandreou had been a graduate student in economics at

Harvard and had taught there and at the University of Minnesota, Northwestern and at the University of California at Berkeley, where he had been chairman of the Department of Economics. Hearing the news from Athens, I was concerned. So were his friends around the country.

My telephone rang throughout the early evening. It was known that I was a friend of Lyndon Johnson's; our differences on Vietnam were not yet public knowledge. Toward midnight I decided I must do something. I got Johnson's leading assistant, Joe Califano, on the line at the White House, and he told me the President, who was getting ready to go to Chancellor Adenauer's funeral, was still entertaining some visitors. He got the details on a sheet of paper and went to see Johnson; I hadn't failed to mention the names and the positions of those who had called. Feeling better, I went to bed.

My telephone rang one last time. It was Nicholas Katzenbach, then the Under Secretary of State. Not concealing his own delight, he read me a message he had just received from the White House: "Call up Ken Galbraith and tell him that I've told those Greek bastards to lay off that son-of-a-bitch—whoever he is."

I got up and called those who had called me, quoting the message. Someone gave it to *Newsweek;* a slightly sanitized version was published in the magazine. A copy of the latter having been left in his jail cell by a visitor, Andreas Papandreou was thus informed of his greatly improved situation. Soon he was released from prison and expelled from Greece. After several years of teaching in Canada, he returned to his homeland, and, for better and sometimes for worse, he dominated Greek politics for years to come.

Earlier, on November 23, 1963, I was, as I've told, in Washington involved in a deeply depressing way with the details of J.F.K.'s funeral. Crossing from the White House to the onetime State Department building next door, I met L.B.J., who was now in his first full day as President. He took my arm and guided me back to the vice-presidential quarters, which he still occupied. There, he told me of his commitment to civil rights legislation, then the nation's most contentious political issue, and to Medicare, Medicaid and constructive social change in general. I expressed my deeper concern as to the position and program of the cold warriors, including on Vietnam. On this I had little response. That half hour showed in a small way what would be the Johnson tragedy. A man with a humane, astute and effective view and agenda on domestic social issues would be destroyed by a foreign and associated military policy on which he lacked experience, interest and self-confidence. These matters he delegated to those whose advice and guidance would end his political career and eventually his life.

I had a closer look at this course of history than most. In December 1963, I made a speech in Washington deploring the persistence of poverty in the great cities, in the hills and hollows of the Appalachian Plateau and in the rural South. The time had come for a positive stand to help the poor. It was one of those rare addresses which produced action. L.B.J. read the speech and summoned me to the White House. The time had, indeed, come, he told me, for a war on poverty; we must frame a specific program. He asked me to join with Sargent Shriver, the head of the Peace Corps, and a few others in doing so.

There is always a danger in claiming too dominant a role in such matters; inevitably, one is better informed on what one has done oneself than on the actions and intentions of others. Nonetheless, then and in ensuing years I saw the Johnson antipoverty program at first hand; when it received legislative sanction, I was formally appointed to its committee of guidance. This continued until my increasingly vocal opposition to the war in Vietnam caused Lyndon Johnson to sever my connection. To make the point, he thoughtfully replaced me with another Harvard economics professor.

The War on Poverty—unlike the highly effective Johnson stand on civil rights—had no clearly visible result. It included some very useful initiatives: the Job Corps, which provided worker training for the young, and the Teachers Corps for the educationally deprived school districts. Also the extremely successful Head Start program for the very young. There were also the Community Action Programs, reflecting a tendency when the course is unclear to give a task back to the localities.

Left uncorrected in the attack on poverty, however, was the greatest defect in the American social structure. Nearly all talk of poverty overlooks this basic fact.

The universal cause of poverty is a shortage of money

among those experiencing it. The obvious—indeed, the only relevant—cure is money, a safety net protecting all from deprivation. A rich country can afford an effective minimum income to keep its citizens above the poverty line. This, along with low-cost housing, is essential. Nothing, however, is more resisted. That there will be abuse by the poor—meaning a resort by some to leisure—must be accepted. For the wives and offspring of the affluent, even for college professors, leisure is tolerable, often good. Never for the poor; they must be forced to work. The War on Poverty did useful things, but it ignored the most fundamental response to the problem.

The war in Vietnam brought to an end Johnson's concern for the poor; to it the principal effort and most of the money had to go. Had one wanted a conservative design for halting unwelcome social action and progress, there could have been no better. It also absorbed the liberal political initiative, which was now spent in bitter opposition to our foreign involvement.

I repeat the basic point: the confidence that Johnson had and that he exuded on domestic social programs deserted him when it came to foreign policy and military venture and adventure. The gap was then filled by the generals and by, as they came to be called, the Cold War liberals. The reaction of the military was perhaps inevitable; they do not minimize their role. Having the weapons, they see solution in their use. When I appealed to L.B.J. against the Vietnam commitment, as initially I did, he told me sharply that I should be grateful for the way he was restraining the generals: "You have no idea what they would be doing if I weren't here to stop them."

The blame for Vietnam, however, rests strongly on the President's civilian advisers. There were two controlling factors. The first was the mentality engendered by the Cold War. This accepted with the power of religious belief that Communism was a unified world force. Long after the split between Moscow and Peking was evident, Secretary of State Dean Rusk was still saying, and with emphasis, that the only difference between the Soviets and the Chinese was in how they planned to destroy democracy. It wasn't capitalism they opposed; it was always democracy. To protect democracy, liberals as well as conservatives would have to respond.

Also important was the peculiar satisfaction that civilians derive from military decision making—from showing, as I have said elsewhere, that although they are scholars and intellectuals, they can be as tough in urging aggressive action as anyone in uniform. This was strongly apparent in Vietnam policy; it was intensified by the fact that some of those involved and influential had been junior officers in World War II. Once serving at the lower levels, they were now in command. Their time had come.

I had had such experience personally but, perhaps fortunately, with reverse effect. In the war years in Washington I was brought into close touch with the military establishment. We were fixing the prices of the products that would go into armaments and other military uses. And in 1945, as already noted, I was a director of the United States Strategic Bombing Survey, investigating the effects of bombing on the German war effort. In both positions, I saw the sometimes vacuous nature of much military decision. So did George Ball, Under Secretary of State under Kennedy and Johnson and my fellow Survey director. We both questioned the oft-cited advantage that the United States was presumed to enjoy because of its near-monopoly of air power.

Ball, as the next chapter will tell, continued his opposition within the administration throughout the 1960s. He was heard but ignored. Earlier on, I was simply excluded. My views on Vietnam being known, the comment was "Ken is not useful."

Until the mid-sixties, as I've said, I continued to see L.B.J. on occasion to make my case. Also Robert McNamara at the Pentagon and McGeorge Bundy in the White House, both of whom would listen to dissent. Eventually, however, I concluded that I was serving my own conscience and not getting any result. One must go public, make the war a major political issue. This I did by giving my support to the presidential campaign of Senator Eugene McCarthy. My differences with L.B.J. were now decisive. I never saw him again.

12

Bowles, Ball, Harriman and the Tyranny of Policy

THREE OF MY closest friends in public life were Chester Bowles, George Ball and Averell Harriman. When Bowles and Ball died, I rendered tribute at their funeral services. At the great New York gathering for Harriman, I was an usher, and I used my small authority to seat one or two of the distinguished foreign-policy figures he most disliked with the household staff. Sadly, they didn't seem to notice.

All three of these men were brought to their roles in foreign policy by two equally important motives. There was, first, their commitment to their country and its relations with the larger world and, second, their boredom with any other occupation or non-occupation. Their choice was made easier because all had either ample, or more than ample, resources in wealth and income. Bowles was the well-endowed co-founder of the great early advertising firm of Benton and Bowles. From this had come the money to sustain him in his wartime service in Washington, where eventually he became price, rationing and rent-control czar; in a term as Governor of Connecticut and one in Congress; and then in the post he twice found truly satisfying, Ambassador to India.

George Ball, comfortably favored by his father with a Standard Oil inheritance, was in escape from a Chicago law practice. He began his government career as a lawyer in the New Deal, went on later to become Under Secretary of State and, very briefly, Ambassador to the United Nations.

Harriman, financially the most impressive case, inherited the great wealth of his father, E. H. Harriman, the railroad tycoon or, in the language of the time, robber baron. The younger Harriman had tried railroad management and also finance, and, in later life, he involved himself in practical politics and became Governor of New York. His real sense of fulfillment, personal and public, came, however, from his World War II experience in Moscow and London, about which there will be further word. Thereafter he never awakened happy in the morning if he didn't face a day of foreign-policy concerns. This weakened his influence more than a little; those who worked with him knew all too well that there was never serious danger that he might resign or go public with his dissent.

It was Harriman's wonderful practice to listen, ask questions, rarely to offer his own program or opinion. All instructing him were impressed by his intelligence, allowing, as it did, for his eager receipt of presumptively wise counsel.

For many years I needed no Washington hotel; I had lodging as a matter of course with the Harrimans, as more than occasionally did my wife, at their handsome house in Georgetown, closely adjacent to official Washington, the State Department and Foggy Bottom.

All three men had a focus of interest that was broadly similar. All, but especially Ball, opposed the Vietnam involvement. All, but especially Bowles, had a humane commitment to the poor lands of the planet. All, but especially Harriman, rejected the views of the Cold War enthusiasts and were for maintaining calm relations with the Soviet Union, while not discounting the problems such relations would involve. All opposed the use of the military alliances with, and weapons support to, the poor, and sometimes the poorest, countries as a presumed defense against Communism and the Red Army.

All three, as did I, came afoul of a major feature of our foreign policy. That is its institutional rigidity, which holds it firmly on course even when it is visibly wrong. So it was on Vietnam, as is now accepted. So on the aforementioned military alliances with the poor lands. So it is or was on such matters as the unnecessary enlargement of NATO or the continuing trade and travel sanctions on Cuba. Or, as this is written, on a sensible response to the more liberal tendencies now evident in Iran. This rigidity, with its strong commitment to error, has prevailed in both past and present times, and it requires a special word, for it particularly enchained the three men here principally under discussion.

*

In the main, public agencies in the United States render defined public services and enforce specific laws. There is latitude for good service and bad, but always within a fully established, reasonably evident framework. That is how the Department of Justice, the Department of Agriculture, the Department of Commerce, the Patent Office, all function.

In contrast, the Department of State administers policies, ones that are not usually specified by law but, instead, are an accepted course of action. In the Cold War years there was little detailed legislation telling what should be done as regards the Soviet Union or in the wider world where Communism was thought a threat. Instead, there was a broad unwritten policy that had evolved into a strong, even militant, design in the earlier Dulles years. To that policy all specific action had to conform and all mental commitment be given. Personal belief was adjusted to policy; were it otherwise, the individual could not serve reliably with the organization or live comfortably with himself or herself.

This is the way it remains on foreign policy generally. The man or woman who fully accepts the policy and has belief therein is known as a good soldier. Or a good policy person. The administration of foreign policy thus comes very close to the enforcement of belief. Such commitment is never total; some of those involved manage to maintain their independence of thought. But in so doing, they risk being seen as unreliable—in the common State Department expression earlier quoted, they are not useful. The usual case, however, is acceptance.

A further consequence is that the requisite belief, once established, cannot easily be changed, perhaps not changed at all. One must be consistent in one's faith. One accepts the policy on Cuba, however improbable or obsolete in practical effect. Nothing better describes the making of American foreign policy since World War II than its controlling rigidity in face of original error or needed adjustment.

The principal influence that shaped foreign policy in this period was, as all know, the Communist countries—the Soviet Union and, in lesser measure, China. This was not implausible as regards the Soviet Union; it was obviously a major force in the world. Less plausible was the excessive reliance that was placed on military weaponry, including, in particular, weapons of mass destruction, which, if used, could end all civilized existence.

Most implausible was the assumption that the Third World—Africa, Asia, Central and South America—was vulnerable to the Soviet (and Chinese) promise of Communism. This resulted in a structure of political and military alliances, including those with the poor countries—SEATO and CENTO and the more affluent of ANZUS. No non-Communist government, however incompetent, dictatorial or corrupt, was thought beyond alliance and our formal support. The CIA was also deployed in aid of any anti- or non-Communist political movement wherever it appeared.

That Communism was impossible in a country that had never experienced capitalism was well outside the limits of the accepted belief. This, by a large margin, was our most grievous error as regards Vietnam, as I persistently argued with no evident success. In the deep jungles and scattered villages of Vietnam or, for that matter, those on the Mekong Delta, Communism as an economic system was totally irrelevant. Socialism may be an alternative to capitalism; it is not to a purely rural economy. No one was ever more convincing on this than Marx. Nor was the prospect for Communism appreciably better in the chaotic turbulence of the Vietnamese cities.

In Washington, nonetheless, there was the strongly developed conception of a successful Communist state in Vietnam, which would become a model in country after country for the rest of Southeast Asia, even extending in some of the mentally more incoherent views to India. This, in everyday reference, was the domino theory. On no matter have so many of presumed intelligence been so wrong. The later history of Vietnam has shown that Communists can govern; it has also shown that they cannot sustain a Communist system.

As I have sufficiently noted, I had had personal experience in that part of the world, and I tediously urged the inapplicability of Communism as an economic and social system. The vision of the future that American officials, not excluding university intellectuals, saw for Communism could not be altered or diminished. (Not even Communists were so persuaded as to their own eventual success.) This was the controlling policy.

That incompetent and corrupt governments in the poor lands could be socially and politically more damaging than any conceivable Communist threat was also in conflict with the established belief. Similarly, it was seen but not accepted that an American presence, especially a military role, would be regarded as a new and unwelcome form of imperialism.

Such was the policy in the State Department and its underlying precepts when John F. Kennedy came to office in 1961. It did not change; given the nature of foreign policy that I have just described, it had a life of its own. Dean Rusk, the incoming Secretary of State but an old foreign-policy hand, could not entertain a thought of change; the worldwide threat of Communism, the worldwide system of military alliances, he accepted as holy writ. The threat of Communism, not the reality of government disorganization and poverty, was the ordained belief.

The defeat and elimination of Communism was not only the central but the total purpose of our foreign policy. This, if less devoutly, new appointees in the White House also believed. Although John Foster Dulles was now dead of cancer, the hard line he had articulated was dominant. So was that of his somewhat dim brother Allen, also an influence on policy, who had been retained as head of the CIA. Thus, the way was open for the comic tragedy of the Bay of Pigs, for those who, during the Missile Crisis, accepted the risk of nuclear exchange and for the greatest of all American foreign-policy disasters, Vietnam. Established policy and resulting belief nurtured all of these, the most serious errors or proposals in the history of American foreign relations.

In the Kennedy years and later, this policy shaped the role of my three closest friends. As new arrivals in the State Department, they were openly resistant to both the policy and the supporting belief. All were in positions of importance. Chester Bowles was the new Under Secretary; George Ball was Under Secretary in charge of economic affairs; Averell Harriman had the unfortunately imprecise title of Ambassador-at-Large.

Chester Bowles was the least inhibited critic of the established belief. He was firmly of the view that poverty and repressive government, not Communism, were the operative threat throughout the Third World. Because of this, economic assistance and support for political and social reform, not arms and military alliances, were the relevant response. He had served Harry Truman as Ambassador to India, and there had had a firsthand view of mass poverty, of possible ameliorating solutions and of Jawaharlal Nehru's detachment from the Cold War beliefs and rhetoric. While in New Delhi, he had made his own support of economic change and improvement fully evident; there were occasions when his coverage by the Indian newspapers exceeded that of Nehru himself.

Tall, solid of build and appearance, Chester Bowles was also relentlessly didactic. He never encountered a compelling fact or had a significant thought that he did not convey to others. No one could have been more strongly motivated to challenge the settled system of belief.

Within a few weeks of Bowles's taking office, there came the Bay of Pigs. He learned of it from outside the prevailing cloak of secrecy, told me and others and left no one who had been so told in doubt as to his opposition. (I sent my objections to Kennedy in an unduly careful manner, since I was not supposed to know. They had no effect.) In the months that followed, Bowles continued to give voice to his break with the accepted policy. Others among the new Kennedy arrivals made their peace. He did not. He argued strongly and openly against the military alliances with the poor countries and for economic aid instead. His opposition to the Bay of Pigs was held against him, for it is the mark of a good policy person that he or she does not speak out after failure, let alone disaster. Among those who accepted the established policy there was full agreement: Bowles had to go.

He was promoted to non-employment in the White House—to no function and a general requirement of silence. Then, when my time in India came to an end, he was dispatched to succeed me. It had been seen in my case that adverse views from there could be controlled.

Bowles's public career was effectively at an end. The price of articulate departure from strict organizational precepts—from official policy—could not have been more clear. The man who almost certainly would have been the chief critic of the emerging Vietnam disaster was safely contained.

George Ball became the new Under Secretary. This, it would soon be seen, was a serious Establishment error.

Prior to his appointment, Ball had not been party to the approved State Department beliefs. His primary interest had been in international trade. Also in European union, the latter being a matter to which, with his good friend and close associate Jean Monnet, he had made a major contribution. The Cold War, the military alliances, the poor countries, had not had his attention. He thus seemed a safe successor to Chester Bowles.

Unfortunately for those accepting or approving him, George Ball was governed by a strong and highly independent intelligence. Shortly after he took office, when the Vietnam war became a central issue, he went into firm opposition. Some of his dissent grew out of the experience with military misjudgment that he and I had shared as fellow directors of the U.S. Strategic Bombing Survey.

In the later Kennedy years, and relentlessly in the time of Lyndon Johnson, Ball made his case against the Vietnam war. This he did in extensive and exceedingly competent memoranda and in White House meetings. Sending forces into the jungles of Vietnam would not bring an end to the conflict; it would be merely the prelude to deeper involvement. The promise by the military that each new commitment of troops, each added bombing operation, would be decisive—would be the last—was, he held, essentially worthless.

Ball was a voice alone. Civilians frequently are more than slightly awed by the military presence, and it is a short step, then, from deference to acceptance. In the Vietnam years, there was the devastating tendency of the civilian officials mentioned before to show that they could be as hard-nosed as any general. They could equally well accept or urge the commitment of troops and the risking of lives. From this tendency, George Ball was singularly exempt.

Because of his World War II experience, he was especially doubtful as to the effect of the bombing of North Vietnam. Here, his opposition ran strongly against accepted policy. The military and, needless to say, the Air Force put their most basic trust in the efficacy of bombing. For those involved, it is clean, detached, hygienic, the very opposite of tedious and filthy ground operations, which are costly in casualties. So the hopeful assertion: let air power be unleashed, and the enemy will suffer and surrender.

Nothing was more profoundly at odds with history. There is grave doubt that the strategic long-range bombing of Germany effectively shortened World War II. That war was won by the troops on the ground. Similarly, total control of the air over North Korea did not prevent our devastating defeat at the Yalu, which left that war to be decided—or, rather, rendered into deadlock—by the ground forces. Out of this experience, with much of which he had had intimate association, came Ball's doubts as to the wisdom or practicality of bombing the vast rural reaches of North Vietnam. This was especially so because they were, above all, agricultural, not military, targets.

On Vietnam, Ball was heard with attention, for he was able, even eloquent, in making his case. He had an air of quiet amusement as he cited the more obvious in fantasy or error. But though heard, he had, as already said, no effect. Perhaps the contrary. Decision has greater virtue and force if taken after there has been eloquent dissent. It was all right to make such statements, proclaim such views, so long as they were not carried to a larger public, and this George did not do. All sides had now been heard; that was reassuring. The policy then prevailed.

Averell Harriman's relation to foreign policy was, to say the least, exceptional. As noted, his career had begun in railroading. With World War II came the task of expediting arms and other wartime needs to Britain and his similar service as American representative in Moscow. Then, as the war came to an end, he served as Ambassador to the Court of St. James's at a time when the name of the occupant of that post was still, unlike now, widely known.

After the years spent in the Soviet Union, he was, as I've said, committed to tolerant relations with that country. He was by no means indifferent to the high regard in which he had been held by the Soviet citizenry, a feeling that stayed with him long after the war. A particular incident occurred in my presence. In 1959, he and I were in

Moscow, along with Marie, his tolerant and wonderfully independent wife. One day we left the shabby room they were occupying at the Hotel National on Red Square, a center for privileged visitors, to tour an automobile plant on the edge of the city. As we came into the huge main room, word passed along the lines and across the machines that Harriman was there. Production stopped as workers left their work, one wave after another. They wanted to see the man who, as an ally, had helped save their country from Hitler. Harriman greeted them warmly. It was hard to imagine that anyone could be more pleased; not least was his personal delight that Marie and I were there to observe this truly proletarian reception.

The tolerance he felt for the Soviet Union made Harriman resistant to the accepted Cold War policy, but, alas, he too was ineffective. On the decisive matter of Vietnam he was fully opposed to our participation, something he made abundantly clear in private statements. In public speeches, however, he espoused the official belief. In State Department meetings he expressed his views but then accepted the policy. In Paris he led the negotiations with the North Vietnamese; in conversation there he told me of his opposition to the Washington hard line, but, nonetheless, he went along.

In these years, apart from Bowles, Ball and Harriman, there were others who did not support the official policy. One was Arthur M. Schlesinger, Jr., President Kennedy's well-loved friend, designated historian and literary guide. Unhappily, he was not in the line of command; his views could be dismissed. And they were.

As already noted, Robert McNamara and McGeorge Bundy were almost alone among the influential members of the administration in making themselves available to talk about the Vietnam disaster. As the years passed, McNamara offered a detailed admission of his own error and culpability. This was an act of conscience and courage. Others lived silently with their past for the rest of their lives, knowing only that in any history of the time they would not escape condemnation. The problem was not lack of intelligence; it was acceptance of policy.

The effective opposition to the Vietnam involvement was ultimately to come not from the inside but from the outside. The decisive figure was not one of my old friends from within but Senator Eugene McCarthy of Minnesota from without. A poet and avowed intellectual, he was an early opponent of the war. With Joseph Rauh, the noted Washington lawyer and political leader, I helped persuade him to run for the Democratic presidential nomination in 1968 on the Vietnam issue. He agreed as to the urgency and announced his intention from our house in Cambridge.

Eugene McCarthy quickly became the country's principal and most influential anti-Vietnam spokesman. I campaigned with and for him in the Democratic primaries in New Hampshire, Wisconsin and California, raised money with surprising ease from the many who were both affluent and opposed to the war. Students and also adults turned from essentially futile demonstration to a voice in visibly effective politics. Against Johnson, McCarthy made a strong showing in the early primaries. I was about to mount the platform to speak for him one evening in California when I was handed a note saying that L.B.J. had withdrawn his candidacy; he would not run again. Some improvisation was called for.

Robert Kennedy entered the race on the same issue. I remained with McCarthy. Kennedy won in California, and the night of his victory I was awakened by a call from Richard Goodwin, initially also a McCarthy supporter, who had gone over to Kennedy when he joined the race. He told me it was time to come aboard. I turned on the television and saw Kennedy shot on the floor of the Ambassador Hotel kitchen. As was McCarthy, I was deeply distraught.

At the vigorously contentious convention in the summer of 1968, I was nominally McCarthy's floor manager —nominally, because it was an occasion without visible management by anyone. To a wholly distracted audience, I seconded his nomination. We were defeated by the regular, superior and politically more disciplined Humphrey forces. However, a small event during those days persuaded me that support for the war, Eugene McCarthy's imminent defeat notwithstanding, was at an end.

The political headquarters of the contenders was, as

at earlier conventions, in the Conrad Hilton Hotel on Michigan Avenue. In the park across the street, the mostly young antiwar protesters gathered by the hundreds. A line of National Guardsmen stood between them and the street to keep them away from the hotel. By arrangement, one or another of the McCarthy supporters took turns going across Michigan Avenue to speak to the assemblage and assure them that we were working inside the convention for their cause. One evening it was my turn. Passing through the line of guardsmen, I came to a small platform that had been rigged up and equipped with a loudspeaker.

I told the assembly of our considerable efforts and ended with a plea for nonviolence. With nonviolent protest the armed cannot contend; responding to violent action, they usually prove superior. I then made a statement that immediately I wished I might recover: "I don't want you fighting with these National Guardsmen here behind me. Remember, they're draft dodgers just like you!" On hearing my words, I realized I should not have said them.

I became even more aware of this as I left the platform to make my way back to the hotel. The sergeant in charge stepped from the ranks and said, "Just a minute, sir. Could I have a word with you?" I paused to receive his rebuke.

He came up beside me and said, "I just want to shake your hand, sir. That's the first nice thing anyone has said about us all day."

It was this incident that told me the war could not go on. At whatever distance in time, it would have to end.

Of all those who opposed the war, Eugene McCarthy was the most effective. He made it a major political issue, as those working from the inside—as for so long had I—did not. For his achievement he is insufficiently remembered, which was partly his own choice. As the war wound down, he left the Senate, abandoned public life and went out of the public arena. I have scarcely seen him since.

13

Sketches on the Larger Screen

THE FIRST IMPORTANT political meeting that I attended must have been in either 1919 or 1920, when I was eleven or twelve. It was held on a farm meadow to the south of London, Ontario, north of Lake Erie. My family and I arrived in our Model T, then three or four years old; many others came by horse and buggy. The speaker was William Lyon Mackenzie King, who had just become leader of the Liberal Party, following the death of Sir Wilfrid Laurier, one of the founding figures of modern Canada. King had been much in the Laurier shadow, but that day, and deservedly, he was well received. Alas, no part of his address survives in my memory; it was, nonetheless, my introduction to Canadian political statesmen.

In later times I continued to see King but did not meet him. He ranks as one of the most important and interesting political leaders of the century; it was his political skill that brought Canada through World War II. The desire of his countrymen to fight again in Europe was not great. There were the terrible memories of the First War, in which Canadian troops had suffered an especially comprehensive slaughter; there was no enthusiasm for another battle of Vimy Ridge. Additionally, the French in Canada were not enthusiastic about once again saving France. King brought all together and, with Churchill and F.D.R., was one of the key figures at the conferences in Quebec and off the coast of Newfoundland that guided the course of the war and defined the purposes of the peace, the Four Freedoms. Of this I was an interested but distant observer. As the head of American price-control operations, I was concerned with the coordination of price and rationing policies between Washington and Ottawa. It was not, as it happened, a deeply difficult problem.

To return to my youth. At the time, I heard many political speeches on Mackenzie King's behalf. In the western part of the county of Elgin, neighboring the one in which I first heard King speak, my father was the unchallenged leader of the Grits, as the Liberals were then called; in the United States he would have been a political boss. The major issue under discussion was free trade; this would allow the much desired sale of local farm products to the United States. The Tory opposition was strongly protectionist; in 1911 it had had a notable election victory with the slogan "No truck or trade with the Yankees."

The speech I best remember, one of which I have often told, was by my father at a farm auction. The farmers thus assembled were a sought-after audience; at election time the auctioneer, however reluctantly, would give way temporarily to a political orator. Mounting a large manure pile that day, my father apologized profusely for speaking from the Tory platform. There was an enthusiastic response. The role of public theater in politics has been with me ever since.

Having done graduate work at Harvard, King had a strong preference for subordinates who were similarly qualified. I knew many of them; with several I maintained a lifelong association. Through these Harvard friends, now become top civil servants, I came to know Lester Bowles Pearson and Pierre Elliott Trudeau, each of whom was to be Prime Minister in the years after King.

Lester Pearson was a highly regarded public official and diplomat, a senior Canadian figure in the creation of the United Nations. He received the Nobel Peace Prize for negotiating the end of the 1956 Arab-Israeli war.

After his Civil Service career, he made the transfer from diplomacy to active politics and was elected to the Canadian Parliament, becoming head of the Liberal Party in 1958 and Prime Minister four years later. He is remembered as one of the best—attractive in appearance, warm in greeting, easy in conversation and an engaging speaker. On domestic policy, with both sense and restraint, he did much to bring Canada into the new world of the welfare state. On foreign policy he followed and went beyond Mackenzie King in making his country a major player on the international scene; he was both sensitive and astute in Canada's major foreign-policy concern, which is dealing with the United States. For him, the latter was not an all-powerful neighbor to which, however reluctantly, Canada must adjust; it was a country one understood, did business with as with a good friend and acted independent from as required. The sought-for result was both the protection of national identity and the enhancement of material advantage.

An interesting example of the latter was the automobile pact between the two countries in the mid-1960s. Canada had (and has) a very considerable automobile industry in Windsor, across the Detroit River from Detroit, and in Oshawa, east of Toronto. It was the product, no one could doubt, of tariff protection, but parts had to be manufactured on a less-than-economical scale. Far better would be a larger, more efficient and cheaper supply in both countries, with easy exchange between them. Not without difficulty and patience, a treaty was negotiated, and the final celebratory gathering of statesmen took place at the Johnson ranch in Texas. The press was present in force. L.B.J., as host, spoke warmly of the northern neighbor, of a world model for peaceful cooperation between national states and of Canada's deservedly distinguished Prime Minister. The latter he then introduced as "the Honorable Drew Pearson." Drew Pearson, the most prominent American columnist of his time, was a frequent and unforgiving critic of Johnson. It was sadly evident who was on the President's mind.

I saw Lester Pearson with some frequency during his

life. When one encountered him, there was, first, a warm, friendly embrace; then came an agreeable exchange on the problems of the day. Our shared interest was most strongly affirmed in the summer of 1963, when I had returned to the United States from my years in India and was briefly back at the White House. Pearson and Kennedy had a meeting in Hyannis Port for general talks, including a discussion of the disputatious problem of air rights. When the Canadian airlines, then mostly Air Canada, asked to fly into the United States—to Florida, Philadelphia, Washington and other places of culture or rest—they were asked what Canada had to offer in return. Alas, the great Canadian cities—Montreal, Toronto, Vancouver—are only a few miles from the border. There was no interest on the part of the American airlines in flying to Hudson Bay. So there was stalemate; a few exceptions apart, Canadian flights ended at airports at the border, as did American flights to the north. The American airlines liked the arrangement; it gave them the Canadian passengers for the longer flights within the United States. There was, however, much changing of planes at Buffalo, Detroit and other near-border airports—a miserable arrangement.

The President and the Prime Minister decided to resolve the matter once and for all. A bilateral commission would be established with authority to negotiate, establish binational routes and set matters right. Kennedy announced, "I will appoint Kenneth Galbraith; he is a former Canadian and will understand the problem."

Pearson replied, "Since he was one of ours, I will appoint him too." Kennedy returned to Washington to tell me with delight that I was the one-man member of a two-man commission.

I shared his pleasure and set to work, holding meetings in Ottawa and Washington and talking with executives of the American carriers and those who made airline policy. I was singled out for special attention when I flew from Boston to Washington.

Representing the interests of the two countries, I negotiated vigorously with myself. In a matter of two or three weeks I had allotted routes: Canadian to Chicago, Los Angeles, Florida and wherever; American to the larger Canadian cities. The long-range jet had changed the situation; the U.S. airlines were no longer enthusiastic about landing in Buffalo just to send their passengers on to Toronto on a different carrier. Only Pan Am made objection; this could be ignored. It was a landmark in international negotiation and agreement. In a couple of weeks I reported to Kennedy on my success. He was outraged. "No one will think you accomplished anything in so short a time. Come back and we will announce it in a couple of months." I left, but he allowed me soon to return. All, including Pearson, were pleased. Inevitably, as time passed, my plan also became restrictive and gave way to an open skies agreement. Nothing is forever.

I had a deep feeling of sorrow when Pearson died, in 1972, at the age of seventy-five. My sense of loss was shared throughout the United States, as it was strongly felt in Canada. There were memorial services, including one, well attended, in Boston, for a truly good and valued neighbor who had gone.

The memory of Lester Pearson's personal warmth and intelligence extended to a large and devoted political following in both countries. Many Americans arriving in Washington do not respond well to the names of its airports—celebrating, as they do, Ronald Reagan and John Foster Dulles, especially Dulles. No Canadian and, indeed, no American, has any negative or unpleasant reaction to the name of the well-traveled Toronto airport, the Lester Pearson.

My association with Pierre Trudeau was also agreeable and rewarding over the years. Succeeding Pearson as Prime Minister in 1968, he was proudly French but strongly committed to a single, unified Canada. The Quebec issue, as it is called, is an indispensable topic for Canadian conversation. If all else fails, it can be taken up with the comforting knowledge that nothing new will be said. Then and now, Trudeau has never had time for such diversion. The unity of Canada was for him already settled; one quickly learned that it was a subject he did not think worth discussing.

More open to consideration was his position on Canada's relations with the United States. This was agreeably diverse—and candid. His devotion to the American social and literary scene and to his American friends was very strong. He once said, or is held to have said, "Better a weekend in New York than a week in Montreal." At the same time, and even more than Pearson, he was committed to Canada's independence of action. Thus, when American recognition of Red China was caught up in the obscene politics of the time—the fear of being thought soft on Communism—he led his country on a sensibly separate course: Canada extended recognition. He would have liked to have a strongly Canadian economic policy; the problem here was that such a policy is extensively affected by cyclical international influences, boom and recession, which limit independent action by any one government. In the recession years of the 1970s, Trudeau, nonetheless, furthered economic and employment policies of distinctly Canadian character, including borrowing as necessary to put people to work. He also turned to wage and price controls when the equally disagreeable alternative was severe wage-price inflation. Of such matters we talked and at length. Pierre Trudeau was, I believe, the only political leader ever to acknowledge openly, perhaps recklessly, a commitment to my economic views.

In the French tradition, Pierre Trudeau's interests were not confined to politics and certainly not to economics. He had, as earlier noted, a lively involvement in literary matters and the arts. There is no question that on any list of the most informed and charming politicians on the American or Canadian political scene in our time, he would be at or near the top.

Gunnar Myrdal, the late great Swedish economist and political philosopher, surveyed the world a number of years ago to see which country, allowing always for difficulties growing out of ethnic differences, geography and economic situation, was the best governed. His choice was Canada. So, one cannot doubt, it would be now, and with special mention for Lester Pearson and Pierre Trudeau.

In the years immediately after World War II, the closest connection between the political leaders of any two countries was that between the men and one or two women who were dominant in the British Labour Party, in and out of power, and their counterparts in the Democratic Party in the United States—those who served Truman and the greater number who worked for Adlai Stevenson and then in support of John F. Kennedy. Prominent on the British side were Hugh Gaitskell, Labour Party leader after Clement Attlee; Prime Minister Harold Wilson; Roy Jenkins, Chancellor of the Exchequer and much else; Aneurin Bevan; Anthony Crosland; Richard Crossman; John Strachey; all of them in high and highly visible positions. Also James Callaghan, the last Labour Prime Minister before the long years of Margaret Thatcher; and Denis Healey and Shirley Williams, both of well-deserved prominence, she in England and later in Cambridge, Massachusetts. With my fellow liberals I had a modest role on the American side.

We met in London for discussions that were centered

all but exclusively on needed action within our countries, not on the relations between the two; we met also in the United States, where our house in Cambridge was often the setting as well as the lodging. (It was not thought good to waste money on hotels.) Most, though not quite all, of the British contingent had been to Oxford or Cambridge; Harvard was prominent among the Americans.

Aligned with those active in the House of Commons and the Labour Party were three highly competent, politically involved and very astute economists: Nicholas Kaldor at Cambridge, Thomas Balogh at Oxford and Eric Roll, for long high in the Civil Service. The careers of all three were a tribute to the openness of British public life. All had come to England in escape from the obscene tyrannies of Eastern Europe in the years following the First World War. All three were Jewish. All rose to the top in British political (and intellectual) life—Balogh and Kaldor from their bases in the universities and Roll through the Civil Service to a major role in the making of British economic policy, notably in administering the Marshall Plan, and then on to a directorship of the Bank of England. All three became life peers in the House of Lords. Balogh and Kaldor died in the early 1990s; as this is written, there has been a major celebration in London for Eric Roll's ninetieth birthday.

All those here mentioned had a close, even intimate, concern for American economic policy, and all provided access for their American counterparts to British policy and intended action. Rarely as between countries has there been a closer and, I cannot but think, more satisfactory interchange.

Among the professional politicians in Britain, I had a rewarding association with Hugh Gaitskell and James Callaghan, a more enduring one with John Strachey and the closest and longest of all with Roy Jenkins.

Gaitskell was an admirably informed and deeply committed leader, and he was universally so regarded. In our many discussions there was no problem on which he did not have a considered opinion. He would have become a very great Prime Minister, but death intervened in 1963, when he was only fifty-six. It was a sad loss for us all.

James Callaghan, the holder of various Cabinet posts, including that of Foreign Secretary, became Prime Minister in succession to the less distinguished Harold Wilson in 1976. He then gave way to Margaret Thatcher in 1979. Relaxed, charming, subject to occasional lapses in judgment, he brought a breath of Irish enchantment to English politics. As were Gaitskell and Attlee, he was committed to a humane conception of the welfare state, to the liberties of the citizen and to moderation in the Cold War. I visited him with pleasure in Downing Street for wide-ranging economic and political discussions. My strongest recollection, however, is of sharing the platform with him at an anniversary commemoration of Adam Smith in the latter's birthplace in Kirkcaldy, Scotland, in 1976.

Adam Smith, the founder of modern economics, the primary force against mercantilist control of government and trade, acute in his identification of the essential functions of the state, critical of the basic tendencies of the joint stock company—the corporation—and a supporter of taxation as necessary including a capital levy, is far too important a figure to be left (as he frequently is) to conservatives, who celebrate his emphasis on the force of pecuniary motivation. Not having read him, they consider him theirs.

Present at Kirkcaldy were many such conservatives, including Professor Friedrich von Hayek, the most influential of the faith and the author of *The Road to Serfdom.* Those urging a more cosmic and valid view of Smith were also there.

Kirkcaldy had become an important manufacturing center for linoleum in modern times, and, regrettably, the process involved gives off a very bad smell. Partly for this reason, those attending the ceremony were housed at St. Andrews, the renowned golfing center some twenty-two miles distant, and taxis were sent from Kirkcaldy to bring us to the sessions. On the first day, as the keynote speaker, I was assigned to ride with Callaghan. Like any Labour politician and certainly any Irish member of his craft, he proceeded immediately to engage in conversation the taxi driver, who was more than slightly aware of the identity of his distinguished passenger.

"I suppose, driver, that here in Kirkcaldy you are pretty proud to be celebrating the Adam Smith anniversary today."

"Sure are, sir. Sure are that."

"You know a good deal abut him in these parts?"

"Just that he was the founder of the Labour Party, sir."

The two leading Labour Party figures with whom I had the closest personal friendship were John (Evelyn John St. Loe) Strachey and Roy Jenkins, later and now Lord Jenkins of Hillhead.

I first met John Strachey in India in 1956. A member of a distinguished British family with Indian connections going back to the earliest days of the Raj, he was then offering guidance on the Second Five Year Plan, which was, as I've told, taking form as a broadly socialist design. Strachey, informed, articulate, attractive, could hardly have been more fully prepared for the task. In his earlier years he had been a Communist, though, as he once told me, for tactical reasons he had never been formally enrolled as a member of the party. It was thought that he could be more effective, more influential, just outside. He broke the party association, as did others, at the time of the Hitler-Stalin nonaggression pact.

I was well acquainted with his earlier views, for, in the 1930s, his book *The Coming Struggle for Power* was the best read Marxian tract of the day. At Harvard, along with a number of my colleagues, I assigned it to my tutorial students, not because I fully accepted it but for the doubts it might engender, the discomfort it might cause. Harvard undergraduates were then the sons of the privileged, and after reading the book, they might be forced to wonder as to the permanence of their good fortune, which was so taken for granted, so enjoyed. It was my special pleasure, as that of others, that Strachey, having served as Food Minister in Clement Attlee's postwar Cabinet, went on to be Secretary of State for War. This brought him to Washington and the Pentagon, where the effect on the local Red hunters of having to do business with someone with Strachey's background was, for the observer, a true delight.

After a sojourn in Calcutta at the then-famous Indian Statistical Institute, Strachey, Catherine Galbraith and I traveled extensively over India from the mid-south to Kashmir, in the far north. It was an engaging and illuminating excursion lasting over several weeks and also, in my case, instilling the doubts I have earlier mentioned as to the relevance for India of any comprehensive socialist design. I still encounter the comment I made at the time that that country is, economically speaking, the world's prime example of functional anarchy.

The journey was a vivid introduction to Indian life and culture and to the often bizarre social and political scene. On one particular day it also admirably illustrated Strachey's earlier belief in the force of capitalist and now American imperialism. In Hyderabad, in central India, we were the guests for lunch of the local Rotary Club. As the members filed in, they were introduced by name and occupation:

"Mr. Strachey, Professor Galbraith, this is Rotarian Desai. Rotarian Desai is our leading jeweler here in Hyderabad."

And so it continued until the fourth or fifth in line came along, a small but stalwart man.

"Mr. Strachey, Professor Galbraith, this is Rotarian Sahgal. Rotarian Sahgal is the head of our Communist Party here in Hyderabad."

Our friendship with John Strachey continued in London and the United States in later years. More than anyone, including the book's publisher, Strachey was responsible for getting attention for *The Affluent Society* in Britain. And no one informed me more fully and more wisely on British politics.

In 1963, when my term as Ambassador came to an end in India, Catherine Galbraith, an uncontrollable traveler, went to Afghanistan on her way home, largely because she hadn't been there before. Tired, I flew to Majorca for a short rest. One day I drove from Palma to Deya to meet with Robert Graves, an admired acquaintance from earlier years. It was, as expected, a most agreeable occasion. After my visit I stopped in a small hotel at the top of the hill, where there was a lovely restaurant with a terrace covered by a ceiling of grapevines. Recognizing me, a solemn, sad-faced, bearded young man asked if he could come over to my table; he was of the local literary community, attracted partly by Graves. I bought him a cup of coffee. He told me how miserable was the life of a writer, and I murmured something to the effect that I enjoyed it.

He responded fiercely, "You really don't know what it's like. Your stuff sells."

I remember this conversation, for it was a sad day for me too. When I got back to my hotel in Palma that evening, I found a call had come, saying that John Strachey was dead. I rearranged my schedule and made my way to London to write a memorial note and join with the many who were in mourning. No visit to England has been quite the same since.

The British political leader with whom I have had the closest friendship since Strachey's death, one that continues, is Roy Harris Jenkins. It has been closer than that with almost any other politician, here or abroad.

We first met in the years following World War II, when, after a wartime career as a code breaker at Bletchley Park, the most secret, most sophisticated and perhaps the most important of all British wartime operations, Jenkins was starting his career in Parliament. In 1955, we went together to a Labour Party conference at Margate. It was for me a fine introduction to British political oratory, which is more sustained and possibly even more predictable than its American counterpart. In later years I saw Jenkins become Minister of Aviation, Home Secretary, Chancellor of the Exchequer and President of the European Commission in Brussels. And the principal founder and head of the Social Democratic Party, which for many years he led from the House of Lords while also serving as the Chancellor of Oxford University. All this is only the beginning. He is also Britain's most distinguished political biographer, celebrated for his books on, among others, Stanley Baldwin, William Ewart Gladstone, and by Americans for his life of Harry S Truman.

In conversation, Roy Jenkins does not appear to be giving the political question under consideration careful thought. That is because he has already given it attention; the answer comes effortlessly and at once. His commitment is to peace, negotiation, practical and humane results, never to ideology. It was his dislike of outworn, ideological dogma that brought about his break with the Labour Party, his formation and leadership of the Social Democrats.

We have been, as I indicated, close personal friends for many years—family visits in England and New England, especially in Vermont, for which Jenkins and his wife, Jennifer, share our special affection. Jennifer Jenkins, a longtime leader of the National Trust, has saved many of Britain's architectural monuments for generations present and yet to come.

There can also be unexpected events in a friendship as long and as intimate as ours with the Jenkinses. In our family we recall one of supreme unimportance. Once, when Roy was at the Exchequer, I went to Israel to give some lectures. Returning home, I left in a camp near Haifa my son James, who was then in school and is now a distinguished professor at the University of Texas. He was to fly back, with a stopover in London. Arriving at Heathrow, he duly reported his intended address in England as 11 Downing Street, where, by arrangement with the Jenkinses, he was to stay. He was promptly taken into custody, an action made more plausible by his highly informal attire. An official arrived and asked him who lived at 11 Downing. Making clear that he should know, Jamie said, "The Chancellor of the Exchequer, of course." The official disappeared. Another looked in, asked questions. The bus left for London. Jamie remained in custody; his baggage continued around on the carousel. Finally, another official of higher station arrived, looked carefully at the name on my son's passport and asked if he was related to Galbraith the economist. Learning that he was, the officer released him without apology.

The question is asked and will continue to be asked as to why Roy Jenkins never became Prime Minister. One reason, previously observed, is that he did not surrender to the political commitment of the Labour Party at the time. But also he did not give up other interests for politics. There was his writing. He also had and has a strong commitment to education and intellectual life in general, as manifested by his long tenure as Chancellor of Oxford University. And finally—a matter of occasional adverse mention—he did not, for political effect, deny himself the full range of life's enjoyments. He likes to live well. There is more to life than being a mere politician.

My list of foreign political leaders could be longer. There was no more progressive tendency after World War II than the close connection between the politically concerned in different countries. My own association extended to leaders in France, Germany, Italy and Japan and, in some measure, the Soviet Union. This outer reach I have publicly urged and personally pursued as an important counterbalance to narrow, all-controlling nationalism, which too often is seen as patriotism. Given the worldwide threat of modern weaponry and military doctrine, it is, with much else, a part of our hope for human survival.

And so I come to the end. Most of those of whom I have told are part of history. There is now a new generation of politicians, some of whom I know but about whom I have little to add. My acquaintance with William Jefferson Clinton does extend over the years; Catherine Galbraith and I were once his guests in Arkansas, and I've visited him in Washington, supported him in two elections. But that is all.

During the week in the autumn of 1998 that the House of Representatives dealt with his proposed impeachment, I was in the hospital with a brief but unpleasant attack of pneumonia. I had a deep sense of sorrow that I was not, instead, in a mental institution. There I would have had a stronger sense of affinity with what was happening in Washington.

My closest friend in the Congress for several years was Albert Gore, Sr. He rallied strongly to my defense when, as wartime price czar, I effectively held prices below what could have been had and enjoyed. But Albert Gore's exceptionally able son I only slightly know. He is of a new world. So with others. Once when I was back for a few days at the University of Cambridge, in England, Tony Blair came for a long day and evening tutorial on economics. Alas, that was my only personal encounter with him.

Thus the march to a new generation. This book, these accounts, belong not to the present but to the reasonably recent past. There is one advantage here. Views, interpretations, may change, not excluding those of the participants. The history itself is graven permanently in stone.